An Introduction to Calculus of Variations

Fourth Edition

Aamer Haque

Copyright ©2020 by Aamer Haque. All rights reserved.

Preface

Calculus of variations deals with computing extreme and stationary values of functionals. A functional maps a function (or several functions) to a real number. In science and engineering, functionals often represent a physical quantity of interest.

The equilibrium displacement of an elastic solid, caused by an applied load, is an example of a function of spatial variables. The total potential energy is a functional with the displacement function as its argument. Of all possible displacement functions, the one which minimizes the energy is the actual displacement.

In classical dynamics, the motion of a system is given by some generalized coordinates and either generalized velocities or momenta. The correct dynamical path of the system is the one which minimizes a functional called action.

The purpose of this book is to provide a mathematical introduction to the calculus of variations. Although comprehensive texts on the subject are available, they frequently delve into an extremely detailed assessment of all outcomes. This book avoids many theoretical complications. Instead, the focus is on the main results needed for applications.

The book begins with a review of important results from single and multivariable calculus. The material chosen from those subjects are the ones most directly related to calculus of variations. The book then covers calculus of variations of functionals of a single function. Functionals containing several functions are also discussed. We limit ourselves to functions of a single independent variable. Finally, the main results applied to classical mechanics are given.

This book is meant as an introduction and is not a treatise. Other books of interest are:

- Calculus of Variations by Gelfand and Fomin [9].
- The Variational Principles of Mechanics by Lanczos [21].
- Mechanics by Landau and Lifshitz [22].

These books may be read simultaneously with this book. This book provides many missing mathematical details ignored by other books.

Who should read this book?

- Applied mathematicians with an interest in applied mechanics.
- Students of applied mechanics who desire a solid mathematical foundation for applications requiring calculus of variations.

Those are the two main categories of expected readers. It is assumed that the reader has mastered advanced calculus. A background in real or functional analysis would be helpful, but the necessary concepts and results are provided.

Contents

I Mathematical Background — 9

1 Single Variable Functions — 11
- 1.1 Limits and Continuity — 11
- 1.2 Differentiation — 13
- 1.3 Integral Theorems — 18
- 1.4 Taylor Series — 19
- 1.5 Local Minimums — 22
- 1.6 Variation Notation — 24

2 Vector Spaces — 25
- 2.1 Vector Space — 25
- 2.2 Inner Product Space — 27
- 2.3 Metric Space — 27
- 2.4 Normed Linear Space — 28
- 2.5 Topology — 28
- 2.6 Functionals — 31
- 2.7 Function Spaces — 35

3 Multivariable Functions — 37
- 3.1 Limits and Continuity — 37
- 3.2 Differentiation — 39
- 3.3 Line Integrals — 51
- 3.4 Taylor Series — 52
- 3.5 Local Minimums — 56
- 3.6 Variation Notation — 58
- 3.7 Constrained Minimization — 59

II Calculus of Variations for Single Functions — 63

4 Formulation — 65
- 4.1 Introduction — 65
- 4.2 General Functionals — 66
- 4.3 Alternate Variational Formulation — 72
- 4.4 Integral Functionals — 74
- 4.5 Local Minimums — 79
- 4.6 Fundamental Lemmas — 81

5 Function with 1st Derivative — 89
- 5.1 Fixed End Points — 89
- 5.2 Special Cases — 100
  - 5.2.1 $F \equiv F(y, y')$ — 100
  - 5.2.2 $F \equiv F(x, y')$ — 101
  - 5.2.3 $F \equiv F(x, y)$ — 101
- 5.3 Shortest Distance Between Two Points — 102
- 5.4 Isoparametric Conditions — 104
- 5.5 Shape of a Hanging Cable — 106
- 5.6 Variable End Points — 108
- 5.7 Elastic Bar — 112

6 Function with 1st and 2nd Derivatives — 115
- 6.1 Fixed End Points — 115
- 6.2 Variable End Points — 118
- 6.3 Cantilever Beam — 122

III Calculus of Variations for Multiple Functions — 125

7 Formulation — 127
- 7.1 Introduction — 127
- 7.2 General Functionals — 128
- 7.3 Alternate Variational Formulation — 134
- 7.4 Integral Functionals — 136
- 7.5 Local Minimums — 141

8 Functions with 1st Derivatives — 143
- 8.1 Fixed End Points — 143
- 8.2 Isoparametric Conditions — 150
- 8.3 Auxiliary Constraints — 152
- 8.4 Geodesics — 156
- 8.5 Variable End Points — 158

9 Functions with 1st and 2nd Derivatives — 161
- 9.1 Fixed End Points 161
- 9.2 Variable End Points 167

IV Classical Mechanics — 171

10 Introduction — 173
- 10.1 Newtonian Dynamics 173
- 10.2 Generalized Coordinates 177

11 Single DOF Systems — 179
- 11.1 Lagrangian Dynamics 179
- 11.2 Conservation of Energy 181
- 11.3 Momentum 182
- 11.4 Hamiltonian Dynamics 183
- 11.5 Integration of the Equation of Motion 186
- 11.6 Mass-Spring System 188
- 11.7 Plane Pendulum 189

12 Multiple DOF Systems — 191
- 12.1 Lagrangian Dynamics 191
- 12.2 Conservation of Energy 194
- 12.3 Momentum 195
- 12.4 Hamiltonian Dynamics 196
- 12.5 Action Revisited 200
- 12.6 Canonical Transformations 203
- 12.7 Poisson Brackets 210
- 12.8 Infinitesimal Canonical Transformations 214
- 12.9 Central Force Problem 220
- 12.10 Small Amplitude Oscillations 226

Part I

Mathematical Background

Chapter 1

Single Variable Functions

1.1 Limits and Continuity

This chapter summarizes some important concepts of calculus for a single real variable. Let f be a real-valued function whose domain is an open or closed interval I. We begin with the formal definition of limit of a function: $f(x) \to L$ as $x \to a$.

Definition 1.1.1 (Limit). A function $f(x)$ has *limit* L as $x \to a$ if for every $\varepsilon > 0$, there exists $\delta > 0$ such that $|f(x) - L| < \varepsilon$ whenever $0 < |x - a| < \delta$.
$$\lim_{x \to a} f(x) = L$$

Notice the point $x = a$ is excluded from consideration. It is not required that $f(x)$ be defined at $x = a$. The concept of limit implies that $f(x)$ is close to L whenever x is close to a. The definition of left hand limit $x \to a^-$ simply requires replacing the domain with $0 < a - x < \delta$. In other words, x is always chosen less than a. Right hand limit $x \to a^+$ uses $0 < x - a < \delta$ and thus x is chosen larger than a.

Continuity at a point requires the value of a function be equal to the limit. Let $B(x_0, \delta)$ denote the open interval centered at x_0 with radius δ. In other words, $x \in B(x_0, \delta)$ if $|x - x_0| < \delta$.

Definition 1.1.2 (Continuous). A function $f(x)$ is *continuous* at x_0 if for every $\varepsilon > 0$, there exists $\delta > 0$ such that $|f(x) - f(x_0)| < \varepsilon$ whenever $x \in B(x_0, \delta)$.

An equivalent definition of continuity for single variable functions is: $f(x) \to f(x_0)$ as $x \to x_0$. A function is continuous on an open interval (a, b) if it is continuous at every point $x \in (a, b)$. Continuity on closed intervals $[a, b]$ requires continuity on (a, b) and continuity at the endpoints: $f(x) \to f(a)$ as $x \to a^+$ and $f(x) \to f(b)$ as $x \to b^-$.

The following definitions and lemmas allow us to compare the relative magnitude of functions near a point.

Definition 1.1.3 (Big O Notation). $f(x)$ is called $O(g(x))$ as $x \to a$ if there exist positive constants M and δ such that $|f(x)| < M |g(x)|$ whenever $0 < |x - a| < \delta$.

We remove the point $x = a$ from direct consideration in order to allow for analysis near singular points. Big O notation is a way of stating that $f(x)$ behaves similar to $g(x)$ near the point a.

Lemma 1.1.1. *Suppose we have*
$$\lim_{x \to a} \left| \frac{f(x)}{g(x)} \right| = L$$
where $0 < L < \infty$ and there exists $\gamma > 0$ such that $g(x) \neq 0$ whenever $0 < |x - a| < \gamma$. Then $f(x)$ is $O(g(x))$ as $x \to a$.

Proof. By the definition of limit, we can find $\delta > 0$ such that $||f(x)/g(x)| - L| < 1$ whenever $0 < |x - a| < \delta < \gamma$. This implies that $0 \leq |f(x)/g(x)| < L + 1$. Multiplying through by $|g(x)|$, we get $|f(x)| < (L+1)|g(x)|$. Thus $f(x)$ is $O(g(x))$ as $x \to a$. $\square$

Example. $\sin(x)$ is $O(x)$ as $x \to 0$. This fact follows from the well-known limit:
$$\lim_{x \to 0} \frac{\sin(x)}{x} = 1$$

Definition 1.1.4 (Little o Notation). $f(x)$ is called $o(g(x))$ as $x \to a$ if for every $\varepsilon > 0$, there exists $\delta > 0$ such that $|f(x)| < \varepsilon |g(x)|$ whenever $0 < |x - a| < \delta$.

Little o implies the function $g(x)$ dominates $f(x)$ near the point a. We say that $f(x)$ is much smaller than $g(x)$ near a.

Lemma 1.1.2. *If we have*
$$\lim_{x \to a} \frac{f(x)}{g(x)} = 0$$
and there exists $\gamma > 0$ such that $g(x) \neq 0$ whenever $0 < |x - a| < \gamma$. Then $f(x) = o(g(x))$ as $x \to a$.

Proof. Let $\varepsilon > 0$. Then by the definition of limit, we can find $\delta > 0$ such that $|f(x)/g(x) - 0| < \varepsilon$ whenever $0 < |x - a| < \delta < \gamma$. Thus $|f(x)| < \varepsilon |g(x)|$ whenever $0 < |x - a| < \delta$. $\square$

Example. $\sin(x^2)$ is $o(x)$ as $x \to 0$. This is seen by computing
$$\lim_{x \to 0} \frac{\sin(x^2)}{x} = \lim_{x \to 0} \frac{x \sin(x^2)}{x^2} = \left[\lim_{x \to 0} x \right] \cdot \left[\lim_{x \to 0} \frac{\sin(x^2)}{x^2} \right] = 0 \cdot 1 = 0$$

1.2 Differentiation

The difference quotient definition of derivative is very inconvenient for multi-variable calculus and calculus of variations. An alternative definition is stated here for single variable functions and the same concept is used throughout this book.

Definition 1.2.1 (1st Derivative). A function $f(x)$ has a *1st derivative* on the (open or closed) interval I if there exists a function $f'(x)$ such that

$$f(x + \Delta x) = f(x) + f'(x)\,\Delta x + \varepsilon\,\Delta x, \qquad \forall x \in I \qquad (1.2.1)$$

where $x + \Delta x \in I$ and $\varepsilon \to 0$ as $\Delta x \to 0$.

$\varepsilon\,\Delta x$ is called the error term and it is clearly $o(\Delta x)$ as $\Delta x \to 0$. We note that ε is a function of x and Δx.

The first derivative provides a local linear approximation of a function. This is seen in figure 1.2.1. Let $y = f(x)$ and $\Delta y = f(x + \Delta x) - f(x)$. We then write

$$\Delta y = f'(x)\,\Delta x + \varepsilon\,\Delta x$$

The differential dy is formulated by setting $dx = \Delta x$ and $dy = f'(x)\,\Delta x$. The equation $dy = \left(\frac{dy}{dx}\right)dx$ is sometimes found in books. When Δx is sufficiently small, we have $\Delta y \approx dy$.

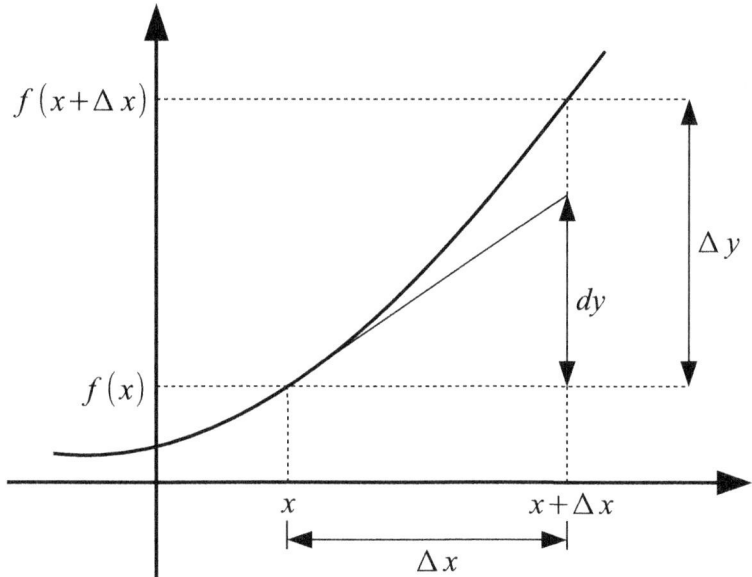

Figure 1.2.1: 1st derivative as linear approximation

The following lemma asserts the equivalence of this definition with the difference quotient definition.

Lemma 1.2.1. *A function $f(x)$ has a 1st derivative on the (open or closed) interval I if and only if*

$$f'(x) = \lim_{\Delta x \to 0} \frac{f(x + \Delta x) - f(x)}{\Delta x}, \qquad \forall x \in I \qquad (1.2.2)$$

where $x + \Delta x \in I$.

Proof. Suppose that equation (1.2.1) defines $f'(x)$. Then we write

$$f'(x) = \frac{f(x + \Delta x) - f(x)}{\Delta x} + \varepsilon$$

Taking the limit of both sides as $\Delta x \to 0$:

$$\lim_{\Delta x \to 0} f'(x) = \lim_{\Delta x \to 0} \left[\frac{f(x + \Delta x) - f(x)}{\Delta x} + \varepsilon \right]$$

The left hand side is simply $f'(x)$. On the right hand side, $\varepsilon \to 0$ as $\Delta x \to 0$. Thus we have equation (1.2.2).

Conversely, suppose that the difference quotient definition (1.2.2) holds. Define ε in the following manner:

$$\varepsilon = \begin{cases} \frac{f(x+\Delta x) - f(x)}{\Delta x} - f'(x) & \text{if } \Delta x \neq 0 \\ 0 & \text{if } \Delta x = 0 \end{cases}$$

Then $\varepsilon \, \Delta x = f(x + \Delta x) - f(x) - f'(x) \, \Delta x$ and we recover equation (1.2.1). □

The uniqueness of the limit in equation (1.2.2) implies that the derivative is unique.

1.2. DIFFERENTIATION

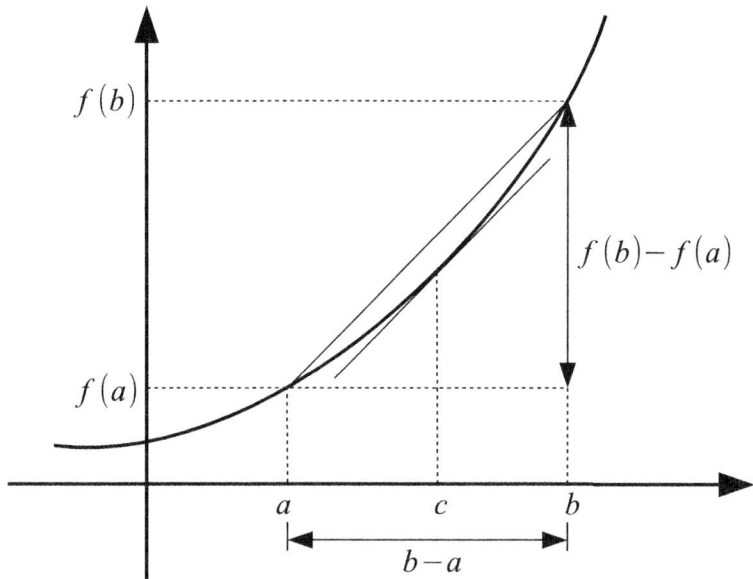

Figure 1.2.2: Differential mean value theorem

The Differential Mean Value Theorem states there is a tangent line with slope equal to that of the secant line connecting two points on the graph of a function. Figure 1.2.2 illustrates this situation. The proof is given in [24, 28, 32].

Theorem 1.2.1 (Differential Mean Value Theorem). *Suppose $f(x)$ is differentiable on the (open or closed) interval I. Then there exists ξ strictly between x and $x + \Delta x$ such that*

$$f'(\xi) = \frac{f(x + \Delta x) - f(x)}{\Delta x}, \qquad \forall x \in I \tag{1.2.3}$$

where $x + \Delta x \in I$. We can write $\xi = x + \theta \Delta x$ where $\theta \in (0, 1)$ and

$$f(x + \Delta x) - f(x) = f'(x + \theta \Delta x) \Delta x, \qquad \forall x \in I \tag{1.2.4}$$

Chain rule of differentiation is proved using definition 1.2.1 and lemma 1.2.1.

Theorem 1.2.2 (Chain Rule)**.** *Suppose $g(x)$ is a differentiable on the (open or closed) interval I and that $f(x)$ is differentiable on the range of $g(x)$. Then the derivative of the composite function $f(g(x))$ is given by*

$$[f(g(x))]' = f'(g(x)) \, g'(x) \qquad (1.2.5)$$

Proof. Let $u = g(x)$ and $\Delta u = g(x + \Delta x) - g(x)$. For all $x \in I$ and $x + \Delta x \in I$, we write

$$\Delta u = g'(x) \, \Delta x + \eta \, \Delta x \qquad (1.2.6)$$

where $\eta \to 0$ as $\Delta x \to 0$.

Let $y = f(u)$ and $\Delta y = f(u + \Delta u) - f(u)$. For all u and $u + \Delta u$ in the range of $g(x)$, we have

$$\Delta y = f'(u) \, \Delta u + \varepsilon \, \Delta u \qquad (1.2.7)$$

where $\varepsilon \to 0$ as $\Delta u \to 0$.

We substitute equation (1.2.6) into equation (1.2.7).

$$\Delta y = f'(g(x)) \, g'(x) \Delta x + [f'(g(x)) \, \eta + \varepsilon \, g'(x) + \varepsilon \eta] \Delta x$$

Forming the difference quotient

$$\frac{\Delta y}{\Delta x} = f'(g(x)) \, g'(x) + [f'(g(x)) \, \eta + \varepsilon \, g'(x) + \varepsilon \eta]$$

We take the limit as $\Delta x \to 0$. We note that $\Delta u \to 0$ as $\Delta x \to 0$ because $g'(x)$ is finite. The term in brackets vanishes and we have

$$[f(g(x))]' = \frac{dy}{dx} = \lim_{\Delta x \to 0} \frac{\Delta y}{\Delta x} = f'(g(x)) \, g'(x)$$

$\square$

Chain rule is often written in the form

$$\frac{dy}{dx} = \frac{dy}{du} \frac{du}{dx}$$

1.2. DIFFERENTIATION

We define higher order derivatives using the next two definitions.

Definition 1.2.2 (2nd Derivative). A function $f(x)$ has a *2nd derivative* on the (open or closed) interval I if it has first derivative $f'(x)$ and there exists a function $f''(x)$ such that

$$f'(x + \Delta x) = f'(x) + f''(x)\Delta x + \varepsilon \Delta x, \qquad \forall x \in I \qquad (1.2.8)$$

where $x + \Delta x \in I$ and $\varepsilon \to 0$ as $\Delta x \to 0$.

The 2nd derivative is simply the derivative of the 1st derivative. Higher order derivatives are defined in a similar manner.

Definition 1.2.3 (Higher Order Derivatives). The n-th order derivative of a function $f(x)$ is the derivative of the $(n-1)$-th order derivative of $f(x)$.

The idea of sets of continuously differentiable functions is necessary for calculus of real functions and calculus of variations.

Definition 1.2.4 (Continuity Class). Let I be an (open or closed) interval in $\mathbb{R}$. $\mathcal{C}^n(I)$ denotes the set of all functions $f : I \to \mathbb{R}$ which have continuous derivatives on I up to order n.

We notate $\mathcal{C}(I) = \mathcal{C}^0(I)$ as the set of continuous functions on I. The definition of continuity class $\mathcal{C}^n$ states that $f \in \mathcal{C}^n$ if and only if $f^{(k)} \in \mathcal{C}(I)$ for $k = 0, \ldots, n$. We observe the proper subset order relation: $\mathcal{C}^{n+1} \subset \mathcal{C}^n$. In other words, there are functions in $\mathcal{C}^n$ which are not in $\mathcal{C}^{n+1}$. An example is provided by $f(x) = |x|$. Clearly $f \in \mathcal{C}^0(\mathbb{R})$ but $f \notin \mathcal{C}^1(\mathbb{R})$.

1.3 Integral Theorems

Some theorems on integrals are required. The proofs are omitted but can be found in [2, 5, 24, 28, 32].

Theorem 1.3.1 (Fundamental Theorem of Calculus). *If $f \in \mathcal{C}(I)$ where I is an (open or closed) interval which contains x_0, then for all $x \in I$ we have*

$$f(x) = \frac{d}{dx} \int_{x_0}^{x} f(s)\, ds$$

$$F(x) - F(x_0) = \int_{x_0}^{x} f(s)\, ds$$

where $F'(x) = f(x)$ on I.

The Integral Mean Value Theorem provides a way of replacing the definite integral of a function by its value at a single point.

Theorem 1.3.2 (Integral Mean Value Theorem). *If $f \in \mathcal{C}(I)$ where I is an (open or closed) interval which contains x_0, then there exists ξ between x_0 and x such that*

$$f(\xi)(x - x_0) = \int_{x_0}^{x} f(s)\, ds, \qquad \forall x \in I \tag{1.3.1}$$

The General Integral Mean Value Theorem is also useful.

Theorem 1.3.3 (General Integral Mean Value Theorem). *If $f, p \in \mathcal{C}(I)$ where I is an (open or closed) interval which contains x_0 and p is non-negative on I, then there exists ξ between x_0 and x such that*

$$f(\xi) \int_{x_0}^{x} p(s)\, ds = \int_{x_0}^{x} f(s)\, p(s)\, ds, \qquad \forall x \in I \tag{1.3.2}$$

The Total Change Theorem is a form of the Fundamental Theorem of Calculus. It states that the total change of a function is the integral of its derivative.

Theorem 1.3.4 (Total Change Theorem). *If $f \in \mathcal{C}^1(I)$ where I is an (open or closed) interval which contains x_0, then for every $x \in I$ we can write*

$$f(x) = f(x_0) + \int_{x_0}^{x} f'(s)\, ds, \qquad \forall x \in I \tag{1.3.3}$$

1.4 Taylor Series

The continuity of the 2nd derivative implies a local quadratic expansion of a function.

Theorem 1.4.1. *If $f \in C^2(I)$ where I is an (open or closed) interval, then*

$$f(x + \Delta x) = f(x) + f'(x)\,\Delta x + \frac{1}{2}f''(x)(\Delta x)^2 + \varepsilon(\Delta x)^2, \qquad \forall x \in I \quad (1.4.1)$$

where $x + \Delta x \in I$ and $\varepsilon \to 0$ as $\Delta x \to 0$.

Proof. We start with the definition of 2nd derivative. For all $x \in I$ with $x + \Delta x \in I$:

$$f'(x + \Delta x) = f'(x) + f''(x)\,\Delta x + \varepsilon_2(x, \Delta x)\,\Delta x$$

where $\varepsilon_2(x, \Delta x) \to 0$ as $\Delta x \to 0$. Notice that we explicitly indicate the dependence of ε_2 on x and Δx. We now integrate the above equation with respect to Δx in the following steps:

$$\int_0^{\Delta x} f'(x + \alpha)\,d\alpha = \int_0^{\Delta x} f'(x)\,d\alpha + \int_0^{\Delta x} f''(x)\,\alpha\,d\alpha + \int_0^{\Delta x} \varepsilon_2(x, \alpha)\,\alpha\,d\alpha$$

$$\int_0^{\Delta x} f'(x + \alpha)\,d\alpha = f'(x)\int_0^{\Delta x} d\alpha + f''(x)\int_0^{\Delta x} \alpha\,d\alpha + \int_0^{\Delta x} \varepsilon_2(x, \alpha)\,\alpha\,d\alpha$$

$$f(x + \Delta x) - f(x) = f'(x)\,\Delta x + \frac{1}{2}f''(x)(\Delta x)^2 + \int_0^{\Delta x} \varepsilon_2(x, \alpha)\,\alpha\,d\alpha$$

The last integral requires further investigation. First, we note that $\varepsilon_2(x, \alpha)$ is a continuous function in both x and α because $f \in C^2(I)$. We next observe that α does not change sign when computing the definite integral. The General Integral Mean Value Theorem can be applied. For the case where $\Delta x \geq 0$,

$$\int_0^{\Delta x} \varepsilon_2(x, \alpha)\,\alpha\,d\alpha = \varepsilon_2(x, \xi^+)\int_0^{\Delta x} \alpha\,d\alpha = \frac{1}{2}\varepsilon_2(x, \xi^+)(\Delta x)^2$$

where $\xi^+ \in [0, \Delta x]$. When $\Delta x < 0$, then $-\alpha > 0$ and

$$\int_0^{\Delta x} \varepsilon_2(x, \alpha)\,\alpha\,d\alpha = \varepsilon_2(x, \xi^-)\int_{\Delta x}^0 (-\alpha)\,d\alpha = \frac{1}{2}\varepsilon_2(x, \xi^-)(\Delta x)^2$$

where $\xi^- \in [\Delta x, 0]$. To complete the proof, we set ε according to

$$\varepsilon = \begin{cases} \frac{1}{2}\varepsilon_2(x, \xi^+) & \text{if } \Delta x \geq 0 \\ \frac{1}{2}\varepsilon_2(x, \xi^-) & \text{if } \Delta x < 0 \end{cases}$$

□

Equation (1.4.1) is useful for proving some theorems for minimization. However, we often require more precise statements concerning the error. The following lemma is used in the proofs of Taylor series expansions.

Lemma 1.4.1. *If $\phi \in C^1([0,1])$, then*

$$\phi(1) = \phi(0) + \int_0^1 \phi'(\alpha)\,d\alpha \tag{1.4.2}$$

If $\phi \in C^2([0,1])$, then we also have

$$\phi(1) = \phi(0) + \phi'(0) + \int_0^1 (1-\alpha)\phi''(\alpha)\,d\alpha \tag{1.4.3}$$

Proof. Equation (1.4.2) is simply the Total Change Theorem. Equation (1.4.3) is proved using integration by parts on the following function:

$$\int_0^1 (1-\alpha)\phi''(\alpha)\,d\alpha = (1-\alpha)\phi'(\alpha)\Big|_0^1 - \int_0^1 (-1)\phi'(\alpha)\,d\alpha$$
$$= -\phi'(0) + \int_0^1 \phi'(\alpha)\,d\alpha$$

Using equation (1.4.2) for the integral on the right hand side we have

$$\int_0^1 (1-\alpha)\phi''(\alpha)\,d\alpha = -\phi'(0) + \phi(1) - \phi(0)$$

Rearranging this equation completes the proof of the lemma. □

Theorem 1.4.2 (0th Order Taylor Series). *If $f \in C^1(I)$ where I is an (open or closed) interval which contains x_0, then for every $x \in I$ we can write $x = x_0 + \Delta x$ and*

$$f(x_0 + \Delta x) = f(x_0) + \Delta x \int_0^1 f'(x_0 + \alpha\,\Delta x)\,d\alpha \tag{1.4.4}$$
$$f(x_0 + \Delta x) = f(x_0) + f'(\xi)\,\Delta x \tag{1.4.5}$$

where ξ is some point between x_0 and x.

Proof. Let $\phi(\alpha) = f(x_0 + \alpha\,\Delta x)$ for $\alpha \in [0,1]$. Clearly $\phi(0) = f(x_0)$ and $\phi(1) = f(x_0 + \Delta x)$. Computing the derivative of $\phi(\alpha)$ by chain rule gives the formula

$$\phi'(\alpha) = f'(x_0 + \alpha\,\Delta x)\,\Delta x$$

Substituting these expressions into equation (1.4.2) proves (1.4.4).

The Integral Mean Value Theorem is applied to equation (1.4.4) to obtain equation (1.4.5). This result is another statement of the Differential Mean Value Theorem. □

1.4. TAYLOR SERIES

Theorem 1.4.3 (1st Order Taylor Series). *If $f \in C^2(I)$ where I is an (open or closed) interval which contains x_0, then for every $x \in I$ we can write $x = x_0 + \Delta x$ and*

$$
\begin{aligned}
f(x_0 + \Delta x) &= f(x_0) + f'(x_0)\,\Delta x \\
&\quad + (\Delta x)^2 \int_0^1 (1-\alpha) f''(x_0 + \alpha\,\Delta x)\,d\alpha \qquad (1.4.6) \\
f(x_0 + \Delta x) &= f(x_0) + f'(x_0)\,\Delta x + \frac{1}{2} f''(\xi)(\Delta x)^2 \qquad (1.4.7)
\end{aligned}
$$

where ξ is some point between x_0 and x.

Proof. As in the proof for the 0th order Taylor series, let $\phi(\alpha) = f(x_0 + \alpha \Delta x)$ for $\alpha \in [0,1]$. We know that $\phi(1) = f(x_0 + \Delta x)$ and $\phi(0) = f(x_0)$. Chain rule gives the following formulas:

$$
\begin{aligned}
\phi'(\alpha) &= f'(x_0 + \alpha\,\Delta x)\,\Delta x \\
\phi''(\alpha) &= f''(x_0 + \alpha\,\Delta x)(\Delta x)^2
\end{aligned}
$$

Substituting these expressions into equation (1.4.3) and noting $\phi'(0) = f'(x_0)\,\Delta x$ proves (1.4.6).

We apply the General Integral Mean Value Theorem to the integral in equation (1.4.6). There exists a $\mu \in [0,1]$ so that

$$
\begin{aligned}
(\Delta x)^2 \int_0^1 (1-\alpha) f''(x_0 + \alpha\,\Delta x)\,d\alpha &= f''(x_0 + \mu\,\Delta x)(\Delta x)^2 \int_0^1 (1-\alpha)\,d\alpha \\
&= f''(\xi)(\Delta x)^2 \int_0^1 (1-\alpha)\,d\alpha
\end{aligned}
$$

where we set $\xi = x_0 + \mu\,\Delta x$. Clearly ξ is between x_0 and x. The integral over α is computed as

$$
\int_0^1 (1-\alpha)\,d\alpha = -\frac{1}{2}(1-\alpha)^2 \Big|_0^1 = \frac{1}{2}
$$

Using this result, we have

$$
(\Delta x)^2 \int_0^1 (1-\alpha) f''(x_0 + \alpha\,\Delta x)\,d\alpha = \frac{1}{2} f''(\xi)(\Delta x)^2
$$

Thus, equation (1.4.7) is proved. $\square$

1.5 Local Minimums

Definition 1.5.1 (Stationary Point). x^* is the location of a *stationary point* of the function $f(x)$ if $f'(x^*) = 0$.

Stationary points are usually called critical numbers in calculus textbooks. For differentiable functions, being a stationary point is a necessary criteria for an extremum. However, it is not a sufficient condition. Further properties of the function are needed to determine if a stationary point is an extremum. For this book, we shall only consider local minimums. Maximum values of functions will not be discussed.

We restrict our attention to local minimums inside open intervals to avoid complications at endpoints.

Definition 1.5.2 (Local Minimum). x^* is the location of a *local minimum* of $f(x)$ if there exists $\delta > 0$ such that $f(x^*) \leq f(x)$ for all $x \in B(x^*, \delta)$.

The next two theorems provide necessary conditions for a point to be a local minimum of a differentiable function. Let I be an open interval.

Theorem 1.5.1 (1st Order Necessary Condition). *If x^* is the location of a local minimum of $f(x)$ and $f \in C^1(I)$, then $f'(x^*) = 0$.*

Proof. The proof is by contradiction. Suppose that $f'(x^*) \neq 0$. According to the definition of 1st derivative, we have $f(x^* + \Delta x) = f(x^*) + f'(x^*)\Delta x + \varepsilon \Delta x$ for all $x^* + \Delta x \in I$. The error term is $o(\Delta x)$ and we can find $\delta > 0$ such that $|\varepsilon \Delta x| < \frac{1}{2}|f'(x^*)||\Delta x|$ when $0 < |\Delta x| < \delta$. In other words, there is a neighborhood about x^* where the size of error term is less than half the size of the 1st derivative term. Let Δx have sign opposite that of $f'(x^*)$ and notice that $f'(x^*)\Delta x + \varepsilon \Delta x < 0$. Set $x = x^* + \Delta x$. If $f'(x^*) > 0$, then $f(x) < f(x^*)$ for all x such that $0 < x^* - x < \delta$. If $f'(x^*) < 0$, then $f(x) < f(x^*)$ for all x such that $0 < x - x^* < \delta$. This contradicts the assumption that x^* is a local minimum. Thus $f'(x^*)$ must be zero. $\square$

1.5. LOCAL MINIMUMS

Theorem 1.5.2 (2nd Order Necessary Condition). *If x^* is the location of a local minimum of $f(x)$ and $f \in \mathcal{C}^2(I)$, then $f'(x^*) = 0$ and $f''(x^*) \geq 0$.*

Proof. The 1st order necessary condition implies $f'(x^*) = 0$. The remainder of the proof is by contradiction. Suppose that $f''(x^*) < 0$. Then the definition of 2nd derivative with $f'(x^*) = 0$ gives $f(x^* + \Delta x) = f(x^*) + \frac{1}{2}f''(x^*)(\Delta x)^2 + \varepsilon(\Delta x)^2$ for all $x^* + \Delta x \in I$. The error term is $o[(\Delta x)^2]$ and we can find $\delta > 0$ such that $|\varepsilon|(\Delta x)^2 < \frac{1}{2}|f''(x^*)|(\Delta x)^2$ when $0 < |\Delta x| < \delta$. This implies that $\frac{1}{2}f''(x^*)(\Delta x)^2 + \varepsilon(\Delta x)^2 < 0$ for a small neighborhood about x^* excluding x^*. Set $x = x^* + \Delta x$ and observe that $f(x) < f(x^*)$ for all x such that $0 < |x - x^*| < \delta$. This contradicts the assumption that x^* is a local minimum. Thus $f''(x^*)$ must be be non-negative. □

The fact that these theorems are not sufficient is given by the example $f(x) = x^3$. We observe that $f'(x) = 0$ and $f''(0) \geq 0$, but 0 is not the location of a local minimum of $f(x)$.

For twice continuously differentiable functions, the next theorem provides a sufficient condition for a local minimum. Again, we consider an open interval I.

Theorem 1.5.3 (2nd Order Sufficient Condition). *If $f \in \mathcal{C}^2(I)$, $f'(x^*) = 0$, and $f''(x^*) > 0$, then x^* is the location of a local minimum of $f(x)$.*

Proof. Since $f \in \mathcal{C}^2(I)$, there exists $\delta > 0$ such that $f''(x) > 0$ for all $x \in B(x^*, \delta) \subset I$. The 1st order Taylor series expansion of $f(x)$ about $x = x^*$ is

$$f(x) = f(x^*) + f'(x^*)\Delta x + \frac{1}{2}f''(\xi)(\Delta x)^2$$

where $x = x^* + \Delta x \in B(x^*, \delta)$ and ξ is located somewhere between x^* and x. Since $f'(x^*) = 0$ and $f''(\xi) > 0$, we clearly have $f(x) \geq f(x^*)$ for all $x \in B(x^*, \delta)$. Thus x^* is the location of a local minimum of $f(x)$. □

1.6 Variation Notation

Although not required for single variable calculus, the concept of variation of a function is introduced in this section. This is done for the sake of consistency with calculus of variations. The variation δx of the independent variable x is simply any arbitrary increment in x. In other words, we set $\delta x = \Delta x$. The concept of variation of a function is developed in this section. Let I be an open interval.

Definition 1.6.1 (Admissible). The variation δx at $x \in I$ is *admissible* if $x + \delta x \in I$.

Definition 1.6.2 (1st Variation). The *1st variation* of a function $f \in C^1(I)$ is defined by $\delta f \equiv f'(x)\,\delta x$ where δx is any admissible variation at $x \in I$.

Definition 1.6.3 (2nd Variation). The *2nd variation* of a function $f \in C^2(I)$ is defined by $\delta^2 f \equiv \frac{1}{2} f''(x)(\delta x)^2$ where δx is any admissible variation at $x \in I$.

Note that δf and $\delta^2 f$ are functions of x and δx. However, we avoid writing these dependencies unless necessary. The definitions and theorems of the previous section can be reformulated in terms of variations.

Definition 1.6.4 (Stationary Point). x^* is the location of a *stationary point* of the function $f(x)$ if $\delta f = 0$ for all admissible variations δx at x^*.

Theorem 1.6.1 (1st Order Necessary Condition). *If x^* is the location of a local minimum of $f(x)$ and $f \in C^1(I)$, then $\delta f = 0$ for all admissible variations δx at x^*.*

Theorem 1.6.2 (2nd Order Necessary Condition). *If x^* is the location of a local minimum of $f(x)$ and $f \in C^2(I)$, then $\delta f = 0$ and $\delta^2 f \geq 0$ for all admissible variations δx at x^*.*

Theorem 1.6.3 (2nd Order Sufficient Condition). *If $f \in C^2(I)$, $\delta f = 0$, and $\delta^2 f > 0$ for all admissible variations δx at x^*, then x^* is the location of a local minimum of $f(x)$.*

Chapter 2

Vector Spaces

2.1 Vector Space

The concept of a vector space is useful in describing multi-dimensional sets. The operations of inner product and norm provide additional structure to a vector space. The definitions in the this chapter are given in abstract form. The abstract definitions are general and apply to both finite and infinite dimensional vector spaces. More detailed information on these topics are found in [17, 18, 19, 23, 27, 29, 31].

A point in $\mathbb{R}^n$ is specified by its n coordinates $(x_1, \ldots, x_n)$. We also consider $\mathbb{R}^n$ as an n-dimensional vector space. Points are now specified as column vectors.

$$\mathbf{x} = \begin{bmatrix} x_1 \\ \vdots \\ x_n \end{bmatrix} = [x_1, \ldots, x_n]^T$$

The choice of either the coordinate $(x_1, \ldots, x_n)$ or vector $[x_1, \ldots, x_n]^T$ representation is a matter of convenience. Both concepts are used in this book.

Let $\mathbf{x}, \mathbf{y} \in \mathbb{R}^n$ and $\alpha \in \mathbb{R}$. Scalar multiplication and vector addition are defined in the usual manner.

$$\alpha \mathbf{x} = \begin{bmatrix} \alpha x_1 \\ \vdots \\ \alpha x_n \end{bmatrix}, \quad \mathbf{x} + \mathbf{y} = \begin{bmatrix} x_1 + y_1 \\ \vdots \\ x_n + y_n \end{bmatrix}$$

The origin or zero vector is given by $\mathbf{0} = [0, \ldots, 0]^T$. $\mathbb{R}^n$ satisfies the definition of a vector space which is given on the next page.

Definition 2.1.1 (Vector Space). The set V with scalar multiplication and vector addition is a *vector space* if it satisfies the properties listed below.

Let $\mathbf{x}, \mathbf{y}, \mathbf{z} \in V$ and $a, b \in S$ where S is a scalar field.

- Closure: $a\mathbf{x} + b\mathbf{y} \in V$
- Commutative: $\mathbf{x} + \mathbf{y} = \mathbf{y} + \mathbf{x}$
- Associative: $(\mathbf{x} + \mathbf{y}) + \mathbf{z} = \mathbf{x} + (\mathbf{y} + \mathbf{z})$
- Zero vector: There exists $\mathbf{0} \in V$ such that $\mathbf{x} + \mathbf{0} = \mathbf{x}$
- Inverses: There exists $-\mathbf{x} \in V$ such that $\mathbf{x} + (-\mathbf{x}) = \mathbf{0}$
- Scalar identity and scalar zero: $1 \cdot \mathbf{x} = \mathbf{x}$ and $0 \cdot \mathbf{x} = \mathbf{0}$
- Compatibility: $a(b\mathbf{x}) = (ab)\mathbf{x}$
- Distributive: $a(\mathbf{x} + \mathbf{y}) = a\mathbf{x} + a\mathbf{y}$ and $(a + b)\mathbf{x} = a\mathbf{x} + b\mathbf{x}$

We shall only use $S = \mathbb{R}$ as the scalar field. A subset of a vector space is called a *subspace* if it is satisfies the properties of a vector space.

Consider the (finite or infinite) set of vectors $\{\mathbf{x}_i\}$. A *linear combination* is a vector $\mathbf{u}$ of the form

$$\mathbf{u} = \sum_i c_i \mathbf{x}_i$$

where $c_i \in \mathbb{R}$. The following definitions concern sets of vectors.

Definition 2.1.2 (Linearly Independent). The vectors $\{\mathbf{x}_i\}$ are *linearly independent* if the only solution to

$$\sum_i c_i \mathbf{x}_i = \mathbf{0}$$

is $c_i = 0$ for all i.

Definition 2.1.3 (Span). The *span* of the vectors $\{\mathbf{x}_i\}$ is the set of all linear combinations of $\{\mathbf{x}_i\}$.

Definition 2.1.4 (Basis). The vectors $\{\mathbf{x}_i\}$ are a basis for the vector space V if they are linearly independent and their span is V.

The standard basis $\{\mathbf{e}_i\}$ for $\mathbb{R}^n$ are vectors with 1 in the i-th position and 0 elsewhere.

2.2 Inner Product Space

Definition 2.2.1 (Inner Product Space). An *inner product space* is a vector space V with an inner product $\langle \cdot, \cdot \rangle : V \times V \to \mathbb{R}$ which satisfies the properties listed below.

Let $\mathbf{x}, \mathbf{y}, \mathbf{z} \in V$ and $\alpha \in \mathbb{R}$.
- Additive: $\langle \mathbf{x} + \mathbf{y}, \mathbf{z} \rangle = \langle \mathbf{x}, \mathbf{z} \rangle + \langle \mathbf{y}, \mathbf{z} \rangle$
- Homogeneous: $\langle \alpha \mathbf{x}, \mathbf{y} \rangle = \alpha \langle \mathbf{x}, \mathbf{y} \rangle$
- Symmetric: $\langle \mathbf{x}, \mathbf{y} \rangle = \langle \mathbf{y}, \mathbf{x} \rangle$
- Positive definite: $\langle \mathbf{x}, \mathbf{x} \rangle > 0$ if $\mathbf{x} \neq \mathbf{0}$ and $\langle \mathbf{x}, \mathbf{0} \rangle = 0$

The inner product for $\mathbb{R}^n$ is defined by

$$\langle \mathbf{x}, \mathbf{y} \rangle = \mathbf{x} \cdot \mathbf{y} = \mathbf{x}^T \mathbf{y} = \sum_{i=1}^n x_i y_i$$

We usually use the dot product or vector transpose notation for $\mathbb{R}^n$.

Vectors $\mathbf{x}$ and $\mathbf{y}$ are *orthogonal* if $\langle \mathbf{x}, \mathbf{y} \rangle = 0$. The set of vectors $\{\mathbf{x}_i\}$ are *mutually orthogonal* if any pair of them are orthogonal. Mutually orthogonal sets are also linearly independent.

Definition 2.2.2 (Orthonormal). The set of vectors $\{\mathbf{x}_i\}$ is *orthonormal* if they satisfy

$$\langle \mathbf{x}_i, \mathbf{x}_j \rangle = \delta_{ij} = \begin{cases} 1 & \text{if } i = j \\ 0 & \text{if } i \neq j \end{cases} \tag{2.2.1}$$

δ_{ij} is known as the *Kronecker delta function*.

Any linearly independent set can be made orthonormal by the Gram–Schmidt procedure. The standard basis $\{\mathbf{e}_i\}$ for $\mathbb{R}^n$ is clearly orthonormal.

2.3 Metric Space

Definition 2.3.1 (Metric Space). A *metric space* is a set V with a metric (i.e. distance function) $d(\cdot, \cdot) : V \times V \to \mathbb{R}$ which satisfies the properties listed below.

Let $\mathbf{x}, \mathbf{y}, \mathbf{z} \in V$.
- Positive definite: $d(\mathbf{x}, \mathbf{y}) \geq 0$ and $d(\mathbf{x}, \mathbf{y}) = 0$ iff $\mathbf{x} = \mathbf{y}$
- Symmetric: $d(\mathbf{x}, \mathbf{y}) = d(\mathbf{y}, \mathbf{x})$
- Triangle inequality: $d(\mathbf{x}, \mathbf{y}) \leq d(\mathbf{x}, \mathbf{z}) + d(\mathbf{z}, \mathbf{y})$

A metric for $\mathbb{R}^n$ is given by $d(\mathbf{x}, \mathbf{y}) = \left[\sum_{i=1}^n (y_i - x_i)^2 \right]^{1/2}$.

2.4 Normed Linear Space

Definition 2.4.1 (Normed Linear Space). A *normed linear space* is a vector space V with a norm $\|\cdot\| : V \to \mathbb{R}$ which satisfies the properties listed below.

Let $\mathbf{x}, \mathbf{y} \in V$ and $\alpha \in \mathbb{R}$.

- Positive definite: $\|\mathbf{x}\| \geq 0$ and $\|\mathbf{x}\| = 0$ iff $\mathbf{x} = \mathbf{0}$
- Homogeneous: $\|\alpha \mathbf{x}\| = |\alpha| \|\mathbf{x}\|$
- Triangle inequality: $\|\mathbf{x} + \mathbf{y}\| \leq \|\mathbf{x}\| + \|\mathbf{y}\|$

A *unit vector* has a norm equal to 1. The distance between two points $\mathbf{x}, \mathbf{y} \in V$ is given by $d(\mathbf{x}, \mathbf{y}) = \|\mathbf{x} - \mathbf{y}\|$. This definition of distance makes the normed linear space a metric space.

The 2-norm is formulated in terms of the inner product

$$\|\mathbf{x}\|_2 = \sqrt{\mathbf{x} \cdot \mathbf{x}} \tag{2.4.1}$$

This norm satisfies the Cauchy-Schwarz inequality.

Theorem 2.4.1 (Cauchy-Schwarz Inequality). *Let V be an inner product space and normed linear space with 2-norm defined by equation (2.4.1).*

$$|\langle \mathbf{x}, \mathbf{y} \rangle| \leq \|\mathbf{x}\|_2 \|\mathbf{y}\|_2, \qquad \forall \mathbf{x}, \mathbf{y} \in V \tag{2.4.2}$$

The proof is given in [27, 29, 31]. There are norms which do not arise from inner products.

2.5 Topology

Open sets define the topological structure of a vector space V. We discuss the topology of normed linear spaces using the vector norm.

Definition 2.5.1 (Open Ball). The *open ball* $B(\mathbf{x}_0, \delta)$ centered at $\mathbf{x}_0$ with radius δ is the set of all points $\mathbf{x} \in V$ such that $\|\mathbf{x} - \mathbf{x}_0\| < \delta$.

With this definition, we can define open sets in V.

Definition 2.5.2 (Open Set). Ω is an *open set* if for every $\mathbf{x} \in \Omega$, there exists $\delta > 0$ such that $B(\mathbf{x}, \delta) \subset \Omega$.

It is obvious that open balls and $\mathbb{R}^n$ are open sets. The complement of a set Ω is defined as $\Omega' = \{\mathbf{x} \in V \mid \mathbf{x} \notin \Omega\}$. Closed sets are the complements of open sets.

2.5. TOPOLOGY

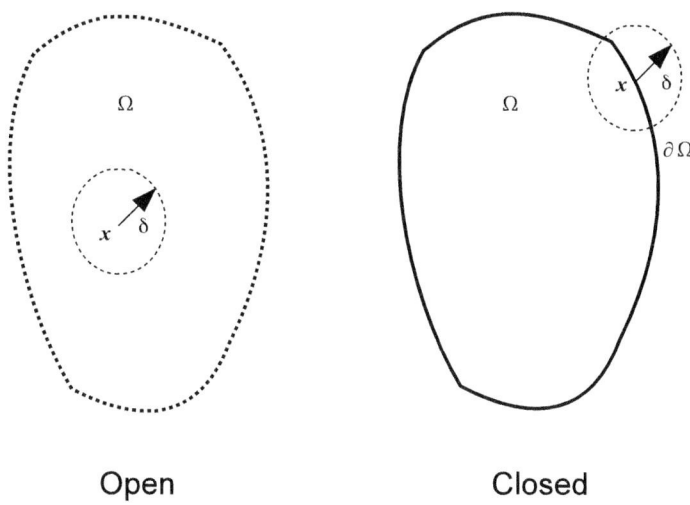

Figure 2.5.1: Open and closed sets

Definition 2.5.3 (Closed Set). Ω is a *closed* set if its complement Ω' is an open set.

Definition 2.5.4 (Boundary). The *boundary* $\partial \Omega$ of an open set Ω is the set of points such that every open ball centered at these points contains at least one point in Ω and one point in Ω'.

The closure of an open set Ω is defined as: $\overline{\Omega} = \Omega \cup \partial \Omega$. The closure $\overline{\Omega}$ is the smallest closed set which contains Ω. Figure 2.5.1 depicts an open set, its boundary, and closure.

Definition 2.5.5 (Bounded Set). A set Ω is *bounded* if there exists a positive constant M such that $\|\mathbf{x}\| < M$ for all $\mathbf{x} \in \Omega$.

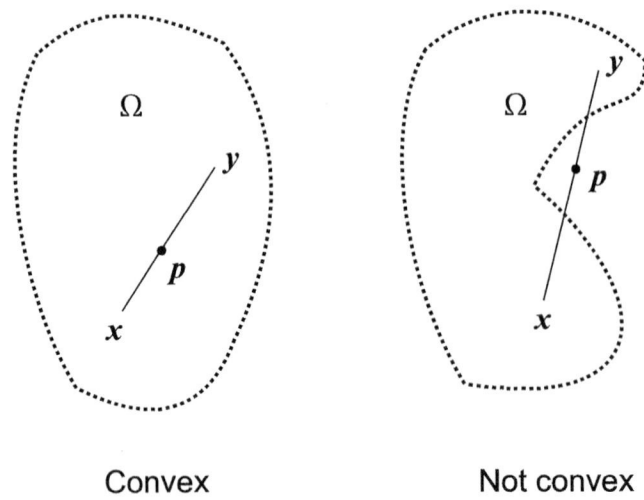

Figure 2.5.2: Convexity of sets

Definition 2.5.6 (Convex Set). A set Ω is *convex* if for all $\mathbf{x}, \mathbf{y} \in \Omega$, the points $\mathbf{p} \in \Omega$ where
$$\mathbf{p} = (1-\lambda)\mathbf{x} + \lambda \mathbf{y}, \qquad \lambda \in [0,1] \tag{2.5.1}$$

Convexity implies that the line connecting any pairs of points in the set remains within the set. Figure 2.5.2 shows examples of a convex and non-convex set. Open balls and $\mathbb{R}^n$ are examples of convex sets. We shall deal extensively with open convex sets in the next chapter.

We also require basic definitions concerning infinite sequences.

Definition 2.5.7 (Cauchy Sequence). A sequence $\{\mathbf{x}_i\}$ is *Cauchy* if for every $\varepsilon > 0$, there exists $N \in \mathbb{N}$ such that $\|\mathbf{x}_i - \mathbf{x}_j\| < \varepsilon$ whenever $i, j > N$.

Definition 2.5.8 (Convergent Sequence). A sequence $\{\mathbf{x}_i\}$ *converges* to $\mathbf{x}$ if for every $\varepsilon > 0$, there exists $N \in \mathbb{N}$ such that $\|\mathbf{x}_i - \mathbf{x}\| < \varepsilon$ whenever $i > N$.

Definition 2.5.9 (Complete Space). A space is *complete* if every Cauchy sequence converges to an element of the space.

Definition 2.5.10 (Banach Space). A complete normed linear space is called a *Banach space*.

$\mathbb{R}^n$ with the 2-norm is a Banach space.

2.6 Functionals

This book deals with functions which map elements of vector spaces to real numbers. Such functions are called functionals. Comprehensive information on functionals is provided in [19, 23, 27]. For finite dimensional vector spaces, further information is found in [8].

Definition 2.6.1 (Continuous Functional). A functional $J : V \to \mathbb{R}$ is *continuous* on the normed linear space V if for every $\varepsilon > 0$, there exists $\delta > 0$ such that $|J(\mathbf{y}) - J(\mathbf{x})| < \varepsilon$ whenever $\|\mathbf{y} - \mathbf{x}\| < \delta$.

Linear functionals play an important role in calculus of variations.

Definition 2.6.2 (Linear Functional). A *linear functional* $L : V \to \mathbb{R}$ on the vector space V satisfies the following properties for all $\mathbf{x}, \mathbf{y} \in V$ and $\alpha \in \mathbb{R}$:

$$L(\alpha \mathbf{x}) = \alpha L(\mathbf{x})$$
$$L(\mathbf{x} + \mathbf{y}) = L(\mathbf{x}) + L(\mathbf{y})$$

The following theorem provides an alternative definition of continuity for linear functionals.

Theorem 2.6.1. *A linear functional $L : V \to \mathbb{R}$ on a normed linear space V is continuous if and only if there exists $M > 0$ such that*

$$|L(\mathbf{x})| \leq M \|\mathbf{x}\|, \qquad \forall \mathbf{x} \in V$$

A proof of this theorem is provided in [27].

All linear functionals on $\mathbb{R}^n$ are represented by inner products.

$$L(\mathbf{x}) = \mathbf{a} \cdot \mathbf{x} = \mathbf{a}^T \mathbf{x}, \qquad \forall \mathbf{x} \in \mathbb{R}^n \tag{2.6.1}$$

for some $\mathbf{a} \in \mathbb{R}^n$. The vector $\mathbf{a}$ defines the linear functional.

The derivative of a functional L with respect to a vector $\mathbf{x}$ is defined as

$$\frac{\partial L}{\partial \mathbf{x}} = \left[\frac{\partial L}{\partial x_1}, \ldots, \frac{\partial L}{\partial x_n} \right]^T \tag{2.6.2}$$

The derivative of the functional (2.6.1) is given by the formula

$$\frac{\partial L}{\partial \mathbf{x}} = \mathbf{a} \tag{2.6.3}$$

Equation (2.6.3) is proved in component form.

$$\left[\frac{\partial L}{\partial \mathbf{x}} \right]_k = \frac{\partial}{\partial x_k} \left[\mathbf{a}^T \mathbf{x} \right] = \frac{\partial}{\partial x_k} \sum_{i=1}^n a_i x_i = \sum_{i=1}^n a_i \frac{\partial x_i}{\partial x_k} = a_k$$

A bilinear functional is linear in each of its arguments.

Definition 2.6.3 (Bilinear Functional). A *bilinear functional* $B : V \times V \to \mathbb{R}$ on the vector space V satisfies the following properties for all $\mathbf{x}, \mathbf{y}, \mathbf{z} \in V$ and $\alpha \in \mathbb{R}$:

$$\begin{aligned} B(\alpha \mathbf{x}, \mathbf{y}) &= \alpha B(\mathbf{x}, \mathbf{y}) \\ B(\mathbf{x} + \mathbf{y}, \mathbf{z}) &= B(\mathbf{x}, \mathbf{z}) + B(\mathbf{y}, \mathbf{z}) \\ B(\mathbf{x}, \alpha \mathbf{y}) &= \alpha B(\mathbf{x}, \mathbf{y}) \\ B(\mathbf{x}, \mathbf{y} + \mathbf{z}) &= B(\mathbf{x}, \mathbf{y}) + B(\mathbf{x}, \mathbf{z}) \end{aligned}$$

A bilinear function is continuous if it is continuous in each of its arguments separately. However, it is more convenient to use an alternative definition.

Definition 2.6.4 (Continuous Bilinear Functional). A bilinear functional $B : V \times V \to \mathbb{R}$ on the normed linear space V is *continuous* if there exists $M > 0$ such that
$$|B(\mathbf{x}, \mathbf{y})| \leq M \|\mathbf{x}\| \|\mathbf{y}\|, \qquad \forall \mathbf{x}, \mathbf{y} \in V$$

Bilinear functionals on $\mathbb{R}^n$ are provided by matrix multiplication.

$$B(\mathbf{x}, \mathbf{y}) = \mathbf{x}^T \mathbf{A} \mathbf{y}, \qquad \forall \mathbf{x}, \mathbf{y} \in \mathbb{R}^n \tag{2.6.4}$$

where $\mathbf{A}$ is the $n \times n$ matrix which defines B.

The derivative of the bilinear functional (2.6.4) with respect to each of its arguments is given by

$$\frac{\partial B}{\partial \mathbf{x}} = \mathbf{A}\mathbf{y}, \qquad \frac{\partial B}{\partial \mathbf{y}} = \mathbf{A}^T \mathbf{x} \tag{2.6.5}$$

The proof of these formulas is done in component form. For the first formula in equation (2.6.5), we have

$$\begin{aligned} \left[\frac{\partial B}{\partial \mathbf{x}}\right]_k &= \frac{\partial}{\partial x_k} \left[\mathbf{x}^T \mathbf{A} \mathbf{y}\right] \\ &= \frac{\partial}{\partial x_k} \sum_{i=1}^n \sum_{j=1}^n A_{ij} x_i y_j \\ &= \sum_{i=1}^n \sum_{j=1}^n A_{ij} y_j \frac{\partial x_i}{\partial x_k} \\ &= \sum_{j=1}^n A_{kj} y_j \\ &= [\mathbf{A}\mathbf{y}]_k \end{aligned}$$

2.6. FUNCTIONALS

The second formula in equation (2.6.5) is derived as

$$\begin{aligned}
\left[\frac{\partial B}{\partial \mathbf{y}}\right]_k &= \frac{\partial}{\partial y_k}\left[\mathbf{x}^T \mathbf{A} \mathbf{y}\right] \\
&= \frac{\partial}{\partial y_k}\sum_{i=1}^{n}\sum_{j=1}^{n} A_{ij} x_i y_j \\
&= \sum_{i=1}^{n}\sum_{j=1}^{n} A_{ij} x_i \frac{\partial y_j}{\partial y_k} \\
&= \sum_{i=1}^{n} A_{ik} x_i \\
&= \sum_{i=1}^{n} A_{ki}^T x_i \\
&= \left[\mathbf{A}^T \mathbf{x}\right]_k
\end{aligned}$$

In this book, we are only interested in bilinear functionals which are symmetric.

Definition 2.6.5 (Symmetric Bilinear Functional). *A bilinear functional $B : V \times V \to \mathbb{R}$ on the vector space V is* symmetric *if $B(\mathbf{x},\mathbf{y}) = B(\mathbf{y},\mathbf{x})$ for all $\mathbf{x},\mathbf{y} \in V$.*

For a symmetric bilinear functional defined on $\mathbb{R}^n$, its associated matrix $\mathbf{A}$ in equation (2.6.4) is symmetric.

Symmetric bilinear functions are often used to define quadratic functionals.

Definition 2.6.6 (Quadratic Functional). *A* quadratic functional *$Q : V \to \mathbb{R}$ on the vector space V is defined by a symmetric bilinear functional $B : V \times V \to \mathbb{R}$ in the following way:*

$$Q(\mathbf{x}) = \frac{1}{2} B(\mathbf{x}, \mathbf{x}), \qquad \forall \mathbf{x} \in V \tag{2.6.6}$$

An important property of quadratic functionals is

$$Q(\alpha \mathbf{x}) = \frac{1}{2} B(\alpha \mathbf{x}, \alpha \mathbf{x}) = \frac{\alpha}{2} B(\mathbf{x}, \alpha \mathbf{x}) = \frac{\alpha^2}{2} B(\mathbf{x}, \mathbf{x}) = \alpha^2 Q(\mathbf{x})$$

for all $\alpha \in \mathbb{R}$ and $\mathbf{x} \in V$.

Consider a bilinear functional on $\mathbb{R}^n$ having the following form:

$$Q(\mathbf{x}) = \frac{1}{2}\mathbf{x}^T \mathbf{A} \mathbf{x} \qquad (2.6.7)$$

The derivative of this functional is

$$\frac{\partial Q}{\partial \mathbf{x}} = \frac{1}{2}\left(\mathbf{A} + \mathbf{A}^T\right)\mathbf{x} \qquad (2.6.8)$$

If $\mathbf{A}$ is symmetric, then we have

$$\frac{\partial Q}{\partial \mathbf{x}} = \mathbf{A}\mathbf{x} \qquad (2.6.9)$$

Equation (2.6.8) is derived in component form.

$$\begin{aligned}
\left[\frac{\partial Q}{\partial \mathbf{x}}\right]_k &= \frac{\partial}{\partial x_k}\left[\frac{1}{2}\mathbf{x}^T \mathbf{A}\mathbf{x}\right] \\
&= \frac{1}{2}\frac{\partial}{\partial x_k}\sum_{i=1}^{n}\sum_{j=1}^{n} A_{ij} x_i x_j \\
&= \frac{1}{2}\sum_{i=1}^{n}\sum_{j=1}^{n} A_{ij} \frac{\partial}{\partial x_k}(x_i x_j) \\
&= \frac{1}{2}\sum_{i=1}^{n}\sum_{j=1}^{n} A_{ij}\left(\frac{\partial x_i}{\partial x_k} x_j + x_i \frac{\partial x_j}{\partial x_k}\right) \\
&= \frac{1}{2}\left[\sum_{j=1}^{n} A_{kj} x_j + \sum_{i=1}^{n} A_{ik} x_i\right] \\
&= \frac{1}{2}\left[\sum_{j=1}^{n} A_{kj} x_j + \sum_{i=1}^{n} A^T_{ki} x_i\right] \\
&= \frac{1}{2}\left\{[\mathbf{A}\mathbf{x}]_k + [\mathbf{A}^T\mathbf{x}]_k\right\} \\
&= \frac{1}{2}\left[\left(\mathbf{A} + \mathbf{A}^T\right)\mathbf{x}\right]_k
\end{aligned}$$

2.7 Function Spaces

Sets of continuous functions are required for calculus of variations. It can be shown that such sets are vector spaces. When equipped with an appropriate norm, they also become Banach spaces. Proofs of these facts are found in [19, 27]. Consider a closed interval $I = [a, b]$.

$\mathcal{C}(I)$ is a Banach space when equipped with the supremum norm $\|\cdot\|_{\sup} : \mathcal{C}(I) \to \mathbb{R}$ defined by

$$\|f\|_{\sup} = \max_{x \in I} |f(x)|, \qquad \forall f \in \mathcal{C}(I) \tag{2.7.1}$$

The supremum norm is always finite for functions in $\mathcal{C}(I)$ because continuous functions on closed intervals obtain a maximum and minimum value.

$\mathcal{C}^n(I)$ is a Banach space when equipped with the norm $\|\cdot\|_{nsup} : \mathcal{C}^n(I) \to \mathbb{R}$ defined by

$$\|f\|_{nsup} = \sum_{k=0}^{n} \left\|f^{(k)}\right\|_{\sup}, \qquad \forall f \in \mathcal{C}^n(I) \tag{2.7.2}$$

Note that the superscript k indicates the order of the derivative of f. The $nsup$ norm is also finite for all functions in $\mathcal{C}^n(I)$ because the derivatives up to order n are continuous on I. An example will clarify the notation. The $1sup$ norm is

$$\|f\|_{1sup} = \sum_{k=0}^{1} \left\|f^{(k)}\right\|_{\sup} = \|f\|_{\sup} + \|f'\|_{\sup}, \qquad \forall f \in \mathcal{C}^1(I)$$

An important property of the Banach space $\mathcal{C}^n(I)$ is convergence to zero in the norm implies convergence to the zero function.

$$\lim_{k \to \infty} \|f_k\|_{nsup} = 0 \quad \Rightarrow \quad \lim_{k \to \infty} f_k(x) = 0, \quad \forall x \in I \tag{2.7.3}$$

Chapter 3

Multivariable Functions

3.1 Limits and Continuity

This chapter provides important results for minimization of multivariable functions. We deal almost exclusively with open convex sets $\Omega \subset \mathbb{R}^n$ and functions $f : \Omega \to \mathbb{R}$.

Definition 3.1.1 (Limit). A function $f(\mathbf{x})$ has *limit* L as $\mathbf{x} \to \mathbf{a}$ if for every $\varepsilon > 0$, there exists $\delta > 0$ such that $|f(\mathbf{x}) - L| < \varepsilon$ whenever $0 < \|\mathbf{x} - \mathbf{a}\|_2 < \delta$.

$$\lim_{\mathbf{x} \to \mathbf{a}} f(\mathbf{x}) = L$$

Limit is a necessary, but insufficient, condition for continuity.

Definition 3.1.2 (Continuous). A function $f(\mathbf{x})$ is *continuous* at $\mathbf{x}_0$ if for every $\varepsilon > 0$, there exists $\delta > 0$ such that $|f(\mathbf{x}) - f(\mathbf{x}_0)| < \varepsilon$ whenever $\mathbf{x} \in B(\mathbf{x}_0, \delta)$.

A function is continuous on open set Ω if it is continuous at every point in Ω.

The multivariable versions of big O and little o are given below. We again remove the point $\mathbf{x} = \mathbf{a}$ from direct consideration in order to allow for analysis near singular points.

Definition 3.1.3 (Big O Notation). $f(\mathbf{x})$ is called $O(g(\mathbf{x}))$ as $\mathbf{x} \to \mathbf{a}$ if there exist positive constants M and δ such that $|f(\mathbf{x})| < M\,|g(\mathbf{x})|$ whenever $0 < \|\mathbf{x} - \mathbf{a}\|_2 < \delta$.

Definition 3.1.4 (Little o Notation). $f(\mathbf{x})$ is called $o(g(\mathbf{x}))$ as $\mathbf{x} \to \mathbf{a}$ if for every $\varepsilon > 0$, there exists $\delta > 0$ such that $|f(\mathbf{x})| < \varepsilon\,|g(\mathbf{x})|$ whenever $0 < \|\mathbf{x} - \mathbf{a}\|_2 < \delta$.

Let p be an integer, then $f(\mathbf{x})$ is $o(\|\mathbf{x}\|_2^p)$ as $\mathbf{x} \to \mathbf{0}$ if

$$\lim_{\mathbf{x} \to \mathbf{0}} \frac{|f(\mathbf{x})|}{\|\mathbf{x}\|_2^p} = 0$$

This criteria is often used as an alternate definition of $o(\|\mathbf{x}\|_2^p)$ as $\mathbf{x} \to \mathbf{0}$.

3.2 Differentiation

Differentiation for multivariable functions is more complicated than for single variable functions. The theory is developed using the concept of linear functionals. We demonstrate that this theory is consistent with the usual formulation in terms of partial derivatives. We first define the derivative of a multivariable function in a particular direction.

Definition 3.2.1 (Directional Derivative). The directional derivative $D_\mathbf{s} f(\mathbf{x})$ in the direction of unit vector $\mathbf{s}$ is computed by

$$D_\mathbf{s} f(\mathbf{x}) = \lim_{\alpha \to 0} \frac{f(\mathbf{x} + \alpha \mathbf{s}) - f(\mathbf{x})}{\alpha} \qquad (3.2.1)$$

The definition assumes that $\mathbf{x} + \alpha \mathbf{s}$ is always in the domain of $f(\mathbf{x})$.

Partial derivatives are defined by computing the directional derivatives along specific coordinate axes. Let $\mathbf{e}_i$ be the unit vector in the direction of the i-th coordinate axis.

Definition 3.2.2 (Partial Derivative). The partial derivative $\frac{\partial}{\partial x_i} f(\mathbf{x})$ of $f(\mathbf{x})$ in the direction of the i-th coordinate axis is

$$\frac{\partial}{\partial x_i} f(\mathbf{x}) = D_{\mathbf{e}_i} f(\mathbf{x}) = \lim_{\alpha \to 0} \frac{f(\mathbf{x} + \alpha \mathbf{e}_i) - f(\mathbf{x})}{\alpha} \qquad (3.2.2)$$

The concept of differentiation of a multivariable function is provided in the following definition.

Definition 3.2.3 (1st Derivative). A function $f(\mathbf{x})$ has a *1st derivative* in the open convex set Ω if for every $\mathbf{x} \in \Omega$, there exists a linear functional $D(\mathbf{x}) : \mathbb{R}^n \to \mathbb{R}$ such that

$$f(\mathbf{x} + \Delta \mathbf{x}) = f(\mathbf{x}) + D(\mathbf{x})[\Delta \mathbf{x}] + r(\Delta \mathbf{x}), \qquad \forall \mathbf{x} \in \Omega \qquad (3.2.3)$$

where $\mathbf{x} + \Delta \mathbf{x} \in \Omega$ and $r(\Delta \mathbf{x})$ is $o(\|\Delta \mathbf{x}\|_2)$ as $\Delta \mathbf{x} \to \mathbf{0}$.

r is called the residual or error function. Notice that the derivative $D(\mathbf{x})$ depends on the location $\mathbf{x}$. The argument of the functional $\Delta \mathbf{x}$ is indicated in the bracket. Since linear functionals on $\mathbb{R}^n$ are inner products, we write

$$D(\mathbf{x})[\mathbf{y}] = \mathbf{d}^T(\mathbf{x}) \mathbf{y}, \qquad \forall \mathbf{x} \in \Omega, \forall \mathbf{y} \in \mathbb{R}^n \qquad (3.2.4)$$

where $\mathbf{d}^T(\mathbf{x}) = [D_1(\mathbf{x}), \ldots, D_n(\mathbf{x})]$ and $\mathbf{y}$ is a column vector.

This page intentionally left blank.

3.2. DIFFERENTIATION

The next two theorems relate the concepts of 1st derivative of a function with its partial derivatives.

Theorem 3.2.1. *If $f(\mathbf{x})$ has a 1st derivative in the open convex set Ω, then its partial derivatives exist and*

$$D(\mathbf{x})[\mathbf{y}] = \left[\frac{\partial}{\partial x_1}f(\mathbf{x}), \ldots, \frac{\partial}{\partial x_n}f(\mathbf{x})\right]^T \mathbf{y}, \qquad \forall \mathbf{x} \in \Omega, \forall \mathbf{y} \in \mathbb{R}^n, \qquad (3.2.5)$$

Proof. For each $i = 1, \ldots, n$ let $\Delta \mathbf{x} = \Delta x_i\, \mathbf{e}_i$ such that $\mathbf{x} + \Delta \mathbf{x} \in \Omega$. Use equation (3.2.4) with the expansion (3.2.3).

$$f(\mathbf{x} + \Delta x_i\, \mathbf{e}_i) = f(\mathbf{x}) + D_i(\mathbf{x})\,\Delta x_i + r(\Delta x_i\, \mathbf{e}_i)$$

where $r(\Delta x_i\, \mathbf{e}_i)$ is $o(|\Delta x_i|)$ as $\Delta x_i \to 0$. Divide each term by Δx_i and solve for $D_i(\mathbf{x})$.

$$D_i(\mathbf{x}) = \frac{f(\mathbf{x} + \Delta x_i\, \mathbf{e}_i) - f(\mathbf{x})}{\Delta x_i} - \frac{r(\Delta x_i\, \mathbf{e}_i)}{\Delta x_i}$$

Take the limit of both sides as $\Delta x_i \to 0$.

$$\lim_{\Delta x_i \to 0} D_i(\mathbf{x}) = \lim_{\Delta x_i \to 0} \left[\frac{f(\mathbf{x} + \Delta x_i\, \mathbf{e}_i) - f(\mathbf{x})}{\Delta x_i} - \frac{r(\Delta x_i\, \mathbf{e}_i)}{\Delta x_i} \right]$$

The left hand side is simply $D_i(\mathbf{x})$. On the right hand side, we observe the definition of partial derivative and remember that $r(\Delta x_i\, \mathbf{e}_i)$ is $o(|\Delta x_i|)$ as $\Delta x_i \to 0$. We now have

$$D_i(\mathbf{x}) = \frac{\partial}{\partial x_i} f(\mathbf{x})$$

□

Unlike the single variable case, the converse of the theorem is not true. Existence of partial derivatives does not guarantee that the function is differentiable. Examples of these situations are found in [6, 32]. Continuity of the partial derivatives is required.

Theorem 3.2.2. *If $f(\mathbf{x})$ has continuous partial derivatives in the open convex set Ω, then $f(\mathbf{x})$ is differentiable in Ω.*

Proof. Choose an $\mathbf{x} \in \Omega$ and let $\varepsilon > 0$. The continuity of partial derivatives implies that there exists $\delta > 0$ such that

$$\left| \frac{\partial}{\partial x_i} f(\mathbf{y}) - \frac{\partial}{\partial x_i} f(\mathbf{x}) \right| < \frac{\varepsilon}{n}, \qquad \forall \mathbf{y} \in B(\mathbf{x}, \delta),\ i = 1, \ldots, n \qquad (3.2.6)$$

δ is chosen small enough so that $B(\mathbf{x}, \delta) \subset \Omega$ because Ω is an open set.

Choose $\Delta \mathbf{x} = [\Delta x_1, \ldots, \Delta x_n]^T$ satisfying $\|\Delta \mathbf{x}\|_2 < \delta$. Define the vectors $\mathbf{v}_i$ as follows:

$$\mathbf{v}_i = \begin{cases} \mathbf{0} & \text{if } i = 0 \\ \sum_{j=1}^{i} \Delta x_j\, \mathbf{e}_j & \text{if } 1 \leq i \leq n \end{cases}$$

We write the telescoping finite sum

$$f(\mathbf{x} + \Delta \mathbf{x}) - f(\mathbf{x}) = \sum_{i=1}^{n} [f(\mathbf{x} + \mathbf{v}_i) - f(\mathbf{x} + \mathbf{v}_{i-1})] \qquad (3.2.7)$$

Since $B(\mathbf{x}, \delta)$ is open and convex, the line segments joining $\mathbf{x} + \mathbf{v}_{i-1}$ and $\mathbf{x} + \mathbf{v}_i$ are contained in $B(\mathbf{x}, \delta)$. Notice that $\mathbf{v}_i = \mathbf{v}_{i-1} + \Delta x_i\, \mathbf{e}_i$. Thus we move along the i-th direction when going from $\mathbf{v}_{i-1}$ to $\mathbf{v}_i$. We apply the Differential Mean Value Theorem to that segment. There exists $\mu_i \in (0, 1)$ such that

$$f(\mathbf{x} + \mathbf{v}_i) - f(\mathbf{x} + \mathbf{v}_{i-1}) = \Delta x_i \frac{\partial}{\partial x_i} f(\mathbf{x} + \mathbf{v}_{i-1} + \mu_i \Delta x_i\, \mathbf{e}_i) \qquad (3.2.8)$$

Equation (3.2.6) implies that

$$\left| \frac{\partial}{\partial x_i} f(\mathbf{x} + \mathbf{v}_{i-1} + \mu_i \Delta x_i\, \mathbf{e}_i) - \frac{\partial}{\partial x_i} f(\mathbf{x}) \right| < \frac{\varepsilon}{n}$$

Multiplying by both sides by $|\Delta x_k|$ gives

$$\left| \Delta x_i \frac{\partial}{\partial x_i} f(\mathbf{x} + \mathbf{v}_{i-1} + \mu_i \Delta x_i\, \mathbf{e}_i) - \Delta x_i \frac{\partial}{\partial x_i} f(\mathbf{x}) \right| \leq \frac{\varepsilon}{n} |\Delta x_i| \qquad (3.2.9)$$

The equality occurs when $\Delta x_i = 0$.

We compute the following absolute error by using equation (3.2.7):

$$\left| f(\mathbf{x} + \Delta \mathbf{x}) - f(\mathbf{x}) - \sum_{i=1}^{n} \Delta x_i \frac{\partial}{\partial x_i} f(\mathbf{x}) \right|$$

$$= \left| \sum_{i=1}^{n} [f(\mathbf{x} + \mathbf{v}_i) - f(\mathbf{x} + \mathbf{v}_{i-1})] - \sum_{i=1}^{n} \Delta x_i \frac{\partial}{\partial x_i} f(\mathbf{x}) \right|$$

$$= \left| \sum_{i=1}^{n} \left[f(\mathbf{x} + \mathbf{v}_i) - f(\mathbf{x} + \mathbf{v}_{i-1}) - \Delta x_i \frac{\partial}{\partial x_i} f(\mathbf{x}) \right] \right|$$

3.2. DIFFERENTIATION

The triangle inequality, equation (3.2.8), and inequality (3.2.9) allows us to write

$$\left| f(\mathbf{x} + \Delta\mathbf{x}) - f(\mathbf{x}) - \sum_{i=1}^{n} \Delta x_i \frac{\partial}{\partial x_i} f(\mathbf{x}) \right|$$

$$= \left| \sum_{i=1}^{n} \left[f(\mathbf{x} + \mathbf{v}_i) - f(\mathbf{x} + \mathbf{v}_{i-1}) - \Delta x_i \frac{\partial}{\partial x_i} f(\mathbf{x}) \right] \right|$$

$$\leq \sum_{i=1}^{n} \left| f(\mathbf{x} + \mathbf{v}_i) - f(\mathbf{x} + \mathbf{v}_{i-1}) - \Delta x_i \frac{\partial}{\partial x_i} f(\mathbf{x}) \right|$$

$$= \sum_{i=1}^{n} \left| \Delta x_i \frac{\partial}{\partial x_i} f(\mathbf{x} + \mathbf{v}_{i-1} + \mu_i \Delta x_i \, \mathbf{e}_i) - \Delta x_i \frac{\partial}{\partial x_i} f(\mathbf{x}) \right|$$

$$\leq \sum_{i=1}^{n} \frac{\varepsilon}{n} |\Delta x_i|$$

Let $M = \max_i |\Delta x_i|$ and notice that

$$\sum_{i=1}^{n} \frac{\varepsilon}{n} |\Delta x_i| = \frac{\varepsilon}{n} \sum_{i=1}^{n} |\Delta x_i| \leq \frac{\varepsilon}{n} \sum_{i=1}^{n} M = \varepsilon M \leq \varepsilon \|\Delta \mathbf{x}\|_2$$

The final desired inequality is

$$\left| f(\mathbf{x} + \Delta\mathbf{x}) - f(\mathbf{x}) - \sum_{i=1}^{n} \Delta x_i \frac{\partial}{\partial x_i} f(\mathbf{x}) \right| \leq \varepsilon \|\Delta \mathbf{x}\|_2$$

This can be written in dot product form.

$$\left| f(\mathbf{x} + \Delta\mathbf{x}) - f(\mathbf{x}) - \mathbf{d}^T(\mathbf{x}) \Delta\mathbf{x} \right| \leq \varepsilon \|\Delta \mathbf{x}\|_2 \quad (3.2.10)$$

where

$$\mathbf{d}^T(\mathbf{x}) = \left[\frac{\partial}{\partial x_1} f(\mathbf{x}), \ldots, \frac{\partial}{\partial x_n} f(\mathbf{x}) \right]$$

The continuity of partial derivatives requires $\varepsilon \to 0$ as $\Delta\mathbf{x} \to 0$. Hence the right hand side of equation (3.2.10) is $o(\|\Delta\mathbf{x}\|_2)$. This implies that $f(\mathbf{x})$ is differentiable with its derivative being the linear functional $D(\mathbf{x})[\Delta\mathbf{x}] = \mathbf{d}^T(\mathbf{x})\Delta\mathbf{x}$. □

We require the concept of continuity of the 1st derivative.

Definition 3.2.4 (Continuous 1st Derivative). The derivative $D(\mathbf{x})$ of $f(\mathbf{x})$ is continuous in the open convex set Ω if for every $\mathbf{x} \in \Omega$ and $\varepsilon > 0$, there exists $\delta > 0$ such that $\|\mathbf{d}(\mathbf{y}) - \mathbf{d}(\mathbf{x})\|_2 < \varepsilon$ whenever $\|\mathbf{y} - \mathbf{x}\|_2 < \delta$ where $\mathbf{y} \in \Omega$.

This definition assumes that the derivative is represented by the vector defined in equation (3.2.4).

Theorem 3.2.3. *If $f(\mathbf{x})$ has a continuous derivative in the open convex set Ω, then the partial derivatives of $f(\mathbf{x})$ are continuous in Ω.*

Proof. Suppose that $f(\mathbf{x})$ has a continuous derivative in Ω. Then for every $\varepsilon > 0$, there exists exists $\delta > 0$ such that $\|\mathbf{d}(\mathbf{y}) - \mathbf{d}(\mathbf{x})\|_2 < \varepsilon$ whenever $\|\mathbf{y} - \mathbf{x}\|_2 < \delta$ where $\mathbf{y} \in \Omega$. The i-th partial derivative satisfies

$$\begin{aligned}\left|\frac{\partial}{\partial x_i} f(\mathbf{y}) - \frac{\partial}{\partial x_i} f(\mathbf{x})\right| &= \left|\mathbf{d}^T(\mathbf{y}) \mathbf{e}_i - \mathbf{d}^T(\mathbf{x}) \mathbf{e}_i\right| \\ &= \left|[\mathbf{d}(\mathbf{y}) - \mathbf{d}(\mathbf{x})]^T \mathbf{e}_i\right| \\ &\leq \|\mathbf{d}(\mathbf{y}) - \mathbf{d}(\mathbf{x})\|_2 \|\mathbf{e}_i\|_2 \end{aligned}$$

The last line is the Cauchy-Schwarz inequality. Since the derivative is continuous and $\mathbf{e}_i$ is a unit vector

$$\left|\frac{\partial}{\partial x_i} f(\mathbf{y}) - \frac{\partial}{\partial x_i} f(\mathbf{x})\right| < \varepsilon$$

Thus the i-th partial is continuous. □

3.2. DIFFERENTIATION

Theorem 3.2.4. *If $f(\mathbf{x})$ has continuous partial derivatives in the open convex set Ω, then the derivative of $f(\mathbf{x})$ is continuous in Ω.*

Proof. Suppose that the partial derivatives are continuous. For every $\varepsilon > 0$, there exists exists $\delta > 0$ such that

$$\left| \frac{\partial}{\partial x_i} f(\mathbf{y}) - \frac{\partial}{\partial x_i} f(\mathbf{x}) \right| < \frac{\varepsilon}{n}$$

whenever $\|\mathbf{y} - \mathbf{x}\|_2 < \delta$ where $\mathbf{y} \in \Omega$. δ is chosen small enough so that the inequality above is satisfied for $i = 1, \ldots, n$. We now compute

$$\begin{aligned}
\|\mathbf{d}(\mathbf{y}) - \mathbf{d}(\mathbf{x})\|_2 &= \left\| \sum_{i=1}^{n} \left[\frac{\partial}{\partial x_i} f(\mathbf{y}) - \frac{\partial}{\partial x_i} f(\mathbf{x}) \right] \mathbf{e}_i \right\|_2 \\
&\leq \sum_{i=1}^{n} \left\| \left[\frac{\partial}{\partial x_i} f(\mathbf{y}) - \frac{\partial}{\partial x_i} f(\mathbf{x}) \right] \mathbf{e}_i \right\|_2 \\
&\leq \sum_{i=1}^{n} \left| \frac{\partial}{\partial x_i} f(\mathbf{y}) - \frac{\partial}{\partial x_i} f(\mathbf{x}) \right| \|\mathbf{e}_i\|_2 \\
&\leq \sum_{i=1}^{n} \left| \frac{\partial}{\partial x_i} f(\mathbf{y}) - \frac{\partial}{\partial x_i} f(\mathbf{x}) \right| \\
&< \sum_{i=1}^{n} \frac{\varepsilon}{n} \\
&< \varepsilon
\end{aligned}$$

This inequality demonstrates that the derivative is continuous and the theorem is proved. □

We have just proved the following theorem.

Theorem 3.2.5. *$f(\mathbf{x})$ has a continuous derivative in the open convex set Ω if and only if the partial derivatives of $f(\mathbf{x})$ are continuous in Ω.*

This theorem allows us to define the continuity class $\mathcal{C}^1$ using partial derivatives.

Definition 3.2.5. *Let Ω be a open convex subset of $\mathbb{R}^n$. $\mathcal{C}^1(\Omega)$ denotes all continuous functions $f : \Omega \to \mathbb{R}$ with continuous partial derivatives on Ω.*

The 1st derivative is more conveniently written in terms of the gradient.

Definition 3.2.6 (Gradient). The gradient $\nabla f(\mathbf{x})$ is the n-dimensional vector formed by

$$\nabla f(\mathbf{x}) = [D_1(\mathbf{x}), \ldots, D_n(\mathbf{x})]^T = \left[\frac{\partial}{\partial x_1} f(\mathbf{x}), \ldots, \frac{\partial}{\partial x_n} f(\mathbf{x})\right]^T \tag{3.2.11}$$

Using the gradient, the definition of derivative (3.2.3) is formulated as

$$f(\mathbf{x} + \Delta \mathbf{x}) = f(\mathbf{x}) + \nabla f(\mathbf{x}) \cdot \Delta \mathbf{x} + r(\Delta \mathbf{x}), \qquad \forall \mathbf{x} \in \Omega \tag{3.2.12}$$

where $\mathbf{x} + \Delta \mathbf{x} \in \Omega$ and $r(\Delta \mathbf{x})$ is $o(\|\Delta \mathbf{x}\|_2)$ as $\Delta \mathbf{x} \to \mathbf{0}$. We frequently use equation (3.2.12) to define the 1st derivative.

The total differential of f is defined by setting $d\mathbf{x} = \Delta \mathbf{x}$ and computing

$$df = \nabla f(\mathbf{x}) \cdot d\mathbf{x} = \sum_{i=1}^{n} \frac{\partial f}{\partial x_i} dx_i \tag{3.2.13}$$

The directional derivative can be computed using the gradient.

Theorem 3.2.6. *The directional derivative $D_\mathbf{s} f(\mathbf{x})$ in the direction of unit vector $\mathbf{s}$ can be computed by using*

$$D_\mathbf{s} f(\mathbf{x}) = \nabla f(\mathbf{x}) \cdot \mathbf{s} \tag{3.2.14}$$

Proof. Using the linear functional definition of derivative, compute the difference quotient (3.2.1).

$$\begin{aligned}
D_\mathbf{s} f(\mathbf{x}) &= \lim_{\alpha \to 0} \frac{f(\mathbf{x} + \alpha \mathbf{s}) - f(\mathbf{x})}{\alpha} \\
&= \lim_{\alpha \to 0} \frac{f(\mathbf{x}) + \nabla f(\mathbf{x}) \cdot \alpha \mathbf{s} + r(\alpha \mathbf{s}) - f(\mathbf{x})}{\alpha} \\
&= \lim_{\alpha \to 0} \frac{\nabla f(\mathbf{x}) \cdot \alpha \mathbf{s} + r(\alpha \mathbf{s})}{\alpha} \\
&= \lim_{\alpha \to 0} \frac{\alpha \nabla f(\mathbf{x}) \cdot \mathbf{s} + r(\alpha \mathbf{s})}{\alpha} \\
&= \lim_{\alpha \to 0} \left[\nabla f(\mathbf{x}) \cdot \mathbf{s} + \frac{r(\alpha \mathbf{s})}{\alpha}\right] \\
&= \nabla f(\mathbf{x}) \cdot \mathbf{s} + \lim_{\alpha \to 0} \frac{r(\alpha \mathbf{s})}{\alpha}
\end{aligned}$$

$\alpha \to 0$ implies that $\alpha \mathbf{s} \to \mathbf{0}$. Recall that $r(\alpha \mathbf{s})$ is $o(\|\alpha \mathbf{s}\|_2)$ as $\alpha \mathbf{s} \to \mathbf{0}$. Since $\mathbf{s}$ is a unit vector, we have $\|\alpha \mathbf{s}\|_2 = |\alpha|$. These facts imply that the remaining limit term vanishes and the proof is complete. $\square$

3.2. DIFFERENTIATION

Consider a function $g : \Omega \to \mathbb{R}$ where $\Omega \subset \mathbb{R}^n$ is an open convex set. Now assume that $g \in \mathcal{C}^1(\Omega)$. Then the range of $g(\mathbf{x})$ is an open set $I \subset \mathbb{R}$. Suppose that $f : I \to \mathbb{R}$ and $f \in \mathcal{C}^1(I)$. Then the composite function $f \circ g : \Omega \to \mathbb{R}$ is continuous and differentiable. The derivative of the composite function $f(g(\mathbf{x}))$ is given by the chain rule.

Theorem 3.2.7 (Chain Rule). *Suppose $g(\mathbf{x})$ is differentiable on the open convex set Ω and that $f(x)$ is differentiable on the range of $g(\mathbf{x})$. Then the derivative of the composite function $f(g(\mathbf{x}))$ is given by*

$$\nabla f(g(\mathbf{x})) = f'(g(\mathbf{x})) \, \nabla g(\mathbf{x}) \tag{3.2.15}$$

Proof. Let $u = g(\mathbf{x})$ and $\Delta u = g(\mathbf{x} + \Delta \mathbf{x}) - g(\mathbf{x})$. For all $\mathbf{x} \in \Omega$ and $\mathbf{x} + \Delta \mathbf{x} \in \Omega$, we write

$$\Delta u = \nabla g(\mathbf{x}) \cdot \Delta \mathbf{x} + r(\Delta \mathbf{x}) \tag{3.2.16}$$

where $r(\Delta \mathbf{x})$ is $o(\|\Delta \mathbf{x}\|_2)$ as $\Delta \mathbf{x} \to \mathbf{0}$.

Let $y = f(u)$ and $\Delta y = f(u + \Delta u) - f(u)$. For all u and $u + \Delta u$ in the range of $g(\mathbf{x})$, we have

$$\Delta y = f'(u) \, \Delta u + \varepsilon \, \Delta u \tag{3.2.17}$$

where $\varepsilon \to 0$ as $\Delta u \to 0$.

Substitute equation (3.2.16) into equation (3.2.17).

$$\begin{aligned}
\Delta y &= f'(u)[\nabla g(\mathbf{x}) \cdot \Delta \mathbf{x} + r(\Delta \mathbf{x})] + \varepsilon[\nabla g(\mathbf{x}) \cdot \Delta \mathbf{x} + r(\Delta \mathbf{x})] \\
f(u + \Delta u) - f(u) &= f'(u) \, \nabla g(\mathbf{x}) \cdot \Delta \mathbf{x} + f'(u) \, r(\Delta \mathbf{x}) + \varepsilon[\nabla g(\mathbf{x}) \cdot \Delta \mathbf{x} + r(\Delta \mathbf{x})] \\
f(u + \Delta u) &= f(u) + f'(u) \, \nabla g(\mathbf{x}) \cdot \Delta \mathbf{x} \\
&\quad + f'(u) \, r(\Delta \mathbf{x}) + \varepsilon[\nabla g(\mathbf{x}) \cdot \Delta \mathbf{x} + r(\Delta \mathbf{x})]
\end{aligned}$$

The above expression is rewritten as

$$\begin{aligned}
f(g(\mathbf{x} + \Delta \mathbf{x})) &= f(g(\mathbf{x})) + f'(g(\mathbf{x})) \, \nabla g(\mathbf{x}) \cdot \Delta \mathbf{x} \\
&\quad + f'(g(\mathbf{x})) \, r(\Delta \mathbf{x}) + \varepsilon[\nabla g(\mathbf{x}) \cdot \Delta \mathbf{x} + r(\Delta \mathbf{x})]
\end{aligned}$$

$\Delta u \to 0$ as $\Delta \mathbf{x} \to \mathbf{0}$ because $\nabla g(\mathbf{x})$ is finite. This implies that $\varepsilon \to 0$. We note that $f'(g(\mathbf{x}))$ is also finite. Hence we can write

$$f(g(\mathbf{x} + \Delta \mathbf{x})) = f(g(\mathbf{x})) + f'(g(\mathbf{x})) \, \nabla g(\mathbf{x}) \cdot \Delta \mathbf{x} + R(\Delta \mathbf{x}) \tag{3.2.18}$$

where $R(\Delta \mathbf{x}) = f'(g(\mathbf{x})) \, r(\Delta \mathbf{x}) + \varepsilon[\nabla g(\mathbf{x}) \cdot \Delta \mathbf{x} + r(\Delta \mathbf{x})]$ is $o(\|\Delta \mathbf{x}\|_2)$ as $\Delta \mathbf{x} \to \mathbf{0}$. Equation (3.2.18) implies that $\nabla f(g(\mathbf{x})) = f'(g(\mathbf{x})) \, \nabla g(\mathbf{x})$. $\square$

Since partial derivatives are multivariable functions, we can compute their partial derivatives to produce the 2nd partial derivatives.

Definition 3.2.7 (2nd Partial Derivative). The 2nd partial derivative $\frac{\partial^2}{\partial x_i \partial x_j} f(\mathbf{x})$ of $f(\mathbf{x})$ is defined as

$$\frac{\partial^2}{\partial x_i \partial x_j} f(\mathbf{x}) = \frac{\partial}{\partial x_i}\left[\frac{\partial}{\partial x_j} f(\mathbf{x})\right] \qquad (3.2.19)$$

When $i = j$, we use the notation $\frac{\partial^2}{\partial x_i^2} f(\mathbf{x})$.

The next theorem states a sufficient condition for interchanging the order of partial differentiation.

Theorem 3.2.8. *Let $f(\mathbf{x})$ be a continuous function on an open convex set Ω. If $f(\mathbf{x})$ has continuous 1st and 2nd partial derivatives on Ω, then we have*

$$\frac{\partial^2}{\partial x_i \partial x_j} f(\mathbf{x}) = \frac{\partial^2}{\partial x_j \partial x_i} f(\mathbf{x}), \qquad \forall \mathbf{x} \in \Omega \qquad (3.2.20)$$

Proof. Let Δx_i and Δx_j be chosen so that $\mathbf{x} + \Delta x_i \, \mathbf{e}_i$, $\mathbf{x} + \Delta x_j \, \mathbf{e}_j$, and $\mathbf{x} + \Delta x_i \, \mathbf{e}_i + \Delta x_j \, \mathbf{e}_j$ are contained in Ω. Form the expression

$$G = f(\mathbf{x} + \Delta x_i \, \mathbf{e}_i + \Delta x_j \, \mathbf{e}_j) - f(\mathbf{x} + \Delta x_i \, \mathbf{e}_i) - f(\mathbf{x} + \Delta x_j \, \mathbf{e}_j) + f(\mathbf{x})$$

Recall that $\mathbf{x} = [x_1, \ldots, x_n]^T$. Define the function $\phi(x_i)$ as follows:

$$\phi(x_i) = f(\mathbf{x} + \Delta x_j \, \mathbf{e}_j) - f(\mathbf{x})$$

The remaining components of $\mathbf{x}$ are considered to be constant parameters. Then G can be written as

$$G = \phi(x_i + \Delta x_i) - \phi(x_i)$$

The Differential Mean Value Theorem is applied to give

$$G = \phi'(x_i + \theta_i \, \Delta x_i) \, \Delta x_i$$

for some $\theta_i \in (0,1)$. The derivative $\phi'(x_i)$ is computed using the partial derivative of f with respect to x_i.

$$\phi'(x_i) = \frac{\partial}{\partial x_i} f(\mathbf{x} + \Delta x_j \, \mathbf{e}_j) - \frac{\partial}{\partial x_i} f(\mathbf{x})$$

This expression can be considered a function of x_j. The Differential Mean Value Theorem is applied along the $\Delta x_j \, \mathbf{e}_j$ direction.

$$\phi'(x_i) = \frac{\partial}{\partial x_j \partial x_i} f(\mathbf{x} + \theta_j \, \Delta x_j \, \mathbf{e}_j) \, \Delta x_j$$

3.2. DIFFERENTIATION

where $\theta_j \in (0,1)$. Using this result, we replace x_i with $x_i + \theta_i \Delta x_i$ and write

$$G = \frac{\partial}{\partial x_j \partial x_i} f(\mathbf{x} + \theta_i \Delta x_i \, \mathbf{e}_i + \theta_j \Delta x_j \, \mathbf{e}_j) \, \Delta x_i \, \Delta x_j \qquad (3.2.21)$$

Define another function $\psi(x_j)$ as

$$\psi(x_j) = f(\mathbf{x} + \Delta x_i \, \mathbf{e}_i) - f(\mathbf{x})$$

The remaining components of $\mathbf{x}$ are held constant constant. G can also be written as

$$G = \psi(x_j + \Delta x_j) - \psi(x_j)$$

The Differential Mean Value Theorem implies

$$G = \psi'(x_j + \mu_j \Delta x_j) \, \Delta x_j$$

for some $\mu_j \in (0,1)$. The derivative $\psi'(x_j)$ is computed using the partial derivative of f with respect to x_j.

$$\psi'(x_j) = \frac{\partial}{\partial x_j} f(\mathbf{x} + \Delta x_i \, \mathbf{e}_i) - \frac{\partial}{\partial x_j} f(\mathbf{x})$$

This expression is considered a function of x_i and the Differential Mean Value Theorem is applied along the $\Delta x_i \, \mathbf{e}_i$ direction.

$$\psi'(x_j) = \frac{\partial}{\partial x_i \partial x_j} f(\mathbf{x} + \mu_i \Delta x_i \, \mathbf{e}_i) \, \Delta x_i$$

where $\mu_i \in (0,1)$. Replacing x_j with $x_j + \mu_j \Delta x_j$ produces

$$G = \frac{\partial}{\partial x_i \partial x_j} f(\mathbf{x} + \mu_i \Delta x_i \, \mathbf{e}_i + \mu_j \Delta x_j \, \mathbf{e}_j) \, \Delta x_i \, \Delta x_j \qquad (3.2.22)$$

W set equations (3.2.21) and (3.2.22) equal and divide by $\Delta x_i \, \Delta x_j$.

$$\frac{\partial}{\partial x_j \partial x_i} f(\mathbf{x} + \theta_i \Delta x_i \, \mathbf{e}_i + \theta_j \Delta x_j \, \mathbf{e}_j)$$
$$= \frac{\partial}{\partial x_i \partial x_j} f(\mathbf{x} + \mu_i \Delta x_i \, \mathbf{e}_i + \mu_j \Delta x_j \, \mathbf{e}_j)$$

Now simultaneously let $\Delta x_i \to 0$ and $\Delta x_j \to 0$. The continuity of the 2nd partial derivatives implies equation (3.2.20). □

For the remainder of this section, we assume that $f(\mathbf{x})$ has continuous partial derivatives up to 2nd order on an open convex set Ω. The 2nd derivative of a multivariable function is defined in terms of a bilinear functional.

Definition 3.2.8 (2nd Derivative). A function $f(\mathbf{x})$ has a *2nd derivative* in the open convex set Ω if for every $\mathbf{x} \in \Omega$, there exists a bilinear functional $D^2(\mathbf{x}) : \mathbb{R}^n \times \mathbb{R}^n \to \mathbb{R}$ such that

$$D(\mathbf{x}+\Delta\mathbf{x})[\mathbf{y}] = D(\mathbf{x})[\mathbf{y}] + D^2(\mathbf{x})[\Delta\mathbf{x},\mathbf{y}] + r(\Delta\mathbf{x},\mathbf{y}), \qquad \forall \mathbf{x} \in \Omega, \forall \mathbf{y} \in \mathbb{R}^n \quad (3.2.23)$$

where $\mathbf{x} + \Delta\mathbf{x} \in \Omega$ and $r(\Delta\mathbf{x})$ is $o(\|\Delta\mathbf{x}\|_2 \|\mathbf{y}\|_2)$ as $\Delta\mathbf{x} \to \mathbf{0}$.

In order to make sense of this definition, we consider the components of $D(\mathbf{x})$. Recall that its i-th component is the i-th partial derivative. Considering the differentiability of the 1st partial derivatives in terms of equation (3.2.12) implies

$$\frac{\partial}{\partial x_i} f(\mathbf{x}+\Delta\mathbf{x}) = \frac{\partial}{\partial x_i} f(\mathbf{x}) + \nabla\left[\frac{\partial}{\partial x_i} f(\mathbf{x})\right] \cdot \Delta\mathbf{x} + r_i(\Delta\mathbf{x})$$

where $r_i(\Delta\mathbf{x})$ is $o(\|\Delta\mathbf{x}\|_2)$ as $\Delta\mathbf{x} \to \mathbf{0}$. Thus, the derivative of the first partial derivative is given by the gradient

$$\nabla\left[\frac{\partial}{\partial x_i} f(\mathbf{x})\right] = \left[\frac{\partial}{\partial x_1}\frac{\partial}{\partial x_i} f(\mathbf{x}), \ldots, \frac{\partial}{\partial x_n}\frac{\partial}{\partial x_i} f(\mathbf{x})\right]^T$$

The continuity of the 2nd partial derivatives guarantees that this gradient is continuous.

Using the dot product definition $D(\mathbf{x})[\mathbf{y}] = \nabla f(\mathbf{x}) \cdot \mathbf{y}$ and the expressions listed above, equation (3.2.23) becomes

$$\sum_{i=1}^{n} \frac{\partial}{\partial x_i} f(\mathbf{x}+\Delta\mathbf{x})\, y_i = \sum_{i=1}^{n} \frac{\partial}{\partial x_i} f(\mathbf{x})\, y_i + \sum_{i=1}^{n}\sum_{j=1}^{n} \frac{\partial^2}{\partial x_j \partial x_i} f(\mathbf{x})\, \Delta x_j\, y_i + \sum_{i=1}^{n} r_i(\Delta\mathbf{x})\, y_i$$

This equation is formulated in matrix/vector form as

$$\nabla f(\mathbf{x}+\Delta\mathbf{x}) \cdot \mathbf{y} = \nabla f(\mathbf{x}) \cdot \mathbf{y} + [\mathbf{B}(\mathbf{x})\,\Delta\mathbf{x}] \cdot \mathbf{y} + \mathbf{r} \cdot \mathbf{y}$$

where the components of $\mathbf{B}$ and $\mathbf{r}$ are

$$\mathbf{B}_{ij}(\mathbf{x}) = \frac{\partial^2}{\partial x_j \partial x_i} f(\mathbf{x}), \qquad \mathbf{r}_i = r_i(\Delta\mathbf{x})$$

The matrix $\mathbf{B}(\mathbf{x})$ is symmetric since the 2nd partial derivatives are assumed to be continuous. We prefer to use the *Hessian matrix* $\mathbf{H}(\mathbf{x}) = \mathbf{B}^T(\mathbf{x})$.

$$\mathbf{H}_{ij}(\mathbf{x}) = \frac{\partial^2}{\partial x_i \partial x_j} f(\mathbf{x})$$

3.3. LINE INTEGRALS

Equation (3.2.23) is formulated as

$$\nabla f(\mathbf{x} + \Delta \mathbf{x}) \cdot \mathbf{y} = \nabla f(\mathbf{x}) \cdot \mathbf{y} + (\Delta \mathbf{x})^T \mathbf{H}(\mathbf{x}) \mathbf{y} + \mathbf{r} \cdot \mathbf{y} \qquad (3.2.24)$$

$D^2(\mathbf{x})[\Delta \mathbf{x}, \mathbf{y}] = (\Delta \mathbf{x})^T \mathbf{H}(\mathbf{x}) \mathbf{y}$ is a symmetric bilinear form because $\mathbf{H}(\mathbf{x})$ is a symmetric matrix. Notice that $\|\mathbf{r}\|_2$ is $o(\|\Delta \mathbf{x}\|_2)$ as $\Delta \mathbf{x} \to \mathbf{0}$. The error term $r(\Delta \mathbf{x}, \mathbf{y}) = \mathbf{r} \cdot \mathbf{y}$ is analyzed in the following manner:

$$\lim_{\Delta \mathbf{x} \to \mathbf{0}} \frac{|\mathbf{r} \cdot \mathbf{y}|}{\|\Delta \mathbf{x}\|_2 \|\mathbf{y}\|_2} \leq \lim_{\Delta \mathbf{x} \to \mathbf{0}} \frac{\|\mathbf{r}\|_2 \|\mathbf{y}\|_2}{\|\Delta \mathbf{x}\|_2 \|\mathbf{y}\|_2} = \lim_{\Delta \mathbf{x} \to \mathbf{0}} \frac{\|\mathbf{r}\|_2}{\|\Delta \mathbf{x}\|_2} = 0$$

Thus $r(\Delta \mathbf{x}, \mathbf{y})$ is $o(\|\Delta \mathbf{x}\|_2 \|\mathbf{y}\|_2)$ as $\Delta \mathbf{x} \to \mathbf{0}$.

The continuity class of 2nd order is defined according to the continuity of the 2nd partial derivatives.

Definition 3.2.9. Let Ω be a open convex subset of $\mathbb{R}^n$. $\mathcal{C}^2(\Omega)$ denotes all continuous functions $f : \Omega \to \mathbb{R}$ with continuous 1st and 2nd partial derivatives on Ω.

3.3 Line Integrals

We will require the use of line integrals to derive the formulas for Taylor series. Let C be any smooth curve with endpoints $\mathbf{a}$ and $\mathbf{b}$. The curve remains inside an open convex set Ω. We denote the line integral of a vector valued function $\mathbf{f} : \mathbb{R}^n \to \mathbb{R}^n$ along C as

$$\int_C \mathbf{f} \cdot d\mathbf{s}$$

The curve C is usually given parametrically as $\mathbf{z} : [\alpha_0, \alpha_1] \to \mathbb{R}^n$ with $\mathbf{z}(\alpha_0) = \mathbf{a}$ and $\mathbf{z}(\alpha_1) = \mathbf{b}$. Using this parameterization, the line integral is written

$$\int_{\alpha_0}^{\alpha_1} \mathbf{f}(\mathbf{z}(\alpha)) \cdot \mathbf{z}'(\alpha) \, d\alpha$$

where $\mathbf{z}(\alpha) = [z_1(\alpha), \ldots, z_n(\alpha)]^T$ and $\mathbf{z}'(\alpha) = [z_1'(\alpha), \ldots, z_n'(\alpha)]^T$.

The next theorem relates the gradient with line integrals.

Theorem 3.3.1. *Let C be a smooth curve contained within an open convex set Ω with endpoints $\mathbf{a}$ and $\mathbf{b}$. If the gradient of $f(\mathbf{x})$ is continuous in Ω, then*

$$\int_C \nabla f(\mathbf{x}) \cdot d\mathbf{s} = f(\mathbf{b}) - f(\mathbf{a}) \qquad (3.3.1)$$

The proof is omitted but is found in many texts on vector calculus.

3.4 Taylor Series

A local quadratic expansion is possible for multivariable functions of class $\mathcal{C}^2$.

Theorem 3.4.1. *If $f \in \mathcal{C}^2(\Omega)$ where Ω is an open convex subset of $\mathbb{R}^n$, then*

$$f(\mathbf{x}+\Delta\mathbf{x}) = f(\mathbf{x}) + \nabla f(\mathbf{x})\cdot\Delta\mathbf{x} + \frac{1}{2}(\Delta\mathbf{x})^T \mathbf{H}(\mathbf{x})\,\Delta\mathbf{x} + r(\Delta\mathbf{x}), \qquad \forall \mathbf{x}\in\Omega \quad (3.4.1)$$

where $\mathbf{x}+\Delta\mathbf{x} \in \Omega$ and $r(\Delta\mathbf{x})$ is $o(\|\Delta\mathbf{x}\|_2^2)$ as $\Delta\mathbf{x}\to\mathbf{0}$.

Proof. We start with equation (3.2.24). For all $\mathbf{x}\in\Omega$ and $\mathbf{x}+\alpha\Delta\mathbf{x}\in\Omega$

$$\nabla f(\mathbf{x}+\alpha\Delta\mathbf{x})\cdot d\mathbf{s} = \nabla f(\mathbf{x})\cdot d\mathbf{s} + (\alpha\Delta\mathbf{x})^T \mathbf{H}(\mathbf{x})\,d\mathbf{s} + \mathbf{r}(\alpha\Delta\mathbf{x})\cdot d\mathbf{s}$$

where $\alpha \in [0,1]$. We compute the line integral of this equation along the line connecting $\mathbf{x}$ and $\mathbf{x}+\Delta\mathbf{x}$. This line is parameterized as $\mathbf{z}(\alpha) = \mathbf{x}+\alpha\Delta\mathbf{x}$. We note that $\mathbf{z}'(\alpha) = \Delta\mathbf{x}$.

$$\int_0^1 \nabla f(\mathbf{x}+\alpha\Delta\mathbf{x})\cdot\Delta\mathbf{x}\,d\alpha = \int_0^1 \nabla f(\mathbf{x})\cdot\Delta\mathbf{x}\,d\alpha + \int_0^1 (\alpha\Delta\mathbf{x})^T \mathbf{H}(\mathbf{x})\,\Delta\mathbf{x}\,d\alpha$$

$$+ \int_0^1 \mathbf{r}(\alpha\Delta\mathbf{x})\cdot\Delta\mathbf{x}\,d\alpha$$

$$f(\mathbf{x}+\Delta\mathbf{x}) - f(\mathbf{x}) = \nabla f(\mathbf{x})\cdot\Delta\mathbf{x}\int_0^1 d\alpha + (\Delta\mathbf{x})^T \mathbf{H}(\mathbf{x})\,\Delta\mathbf{x}\int_0^1 \alpha\,d\alpha$$

$$+ \int_0^1 \mathbf{r}(\alpha\Delta\mathbf{x})\cdot\Delta\mathbf{x}\,d\alpha$$

$$= \nabla f(\mathbf{x})\cdot\Delta\mathbf{x} + \frac{1}{2}(\Delta\mathbf{x})^T \mathbf{H}(\mathbf{x})\,\Delta\mathbf{x}$$

$$+ \int_0^1 \mathbf{r}(\alpha\Delta\mathbf{x})\cdot\Delta\mathbf{x}\,d\alpha$$

We investigate the error term by first noting the following inequalities:

$$\left|\int_0^1 \mathbf{r}(\alpha\Delta\mathbf{x})\cdot\Delta\mathbf{x}\,d\alpha\right| \le \int_0^1 |\mathbf{r}(\alpha\Delta\mathbf{x})\cdot\Delta\mathbf{x}|\,d\alpha \le \int_0^1 \|\mathbf{r}(\alpha\Delta\mathbf{x})\|_2 \|\Delta\mathbf{x}\|_2\,d\alpha$$

Let $\alpha_{\max} \in [0,1]$ be the value of α which maximizes $\|\mathbf{r}(\alpha\Delta\mathbf{x})\|_2$. We now have

$$\left|\int_0^1 \mathbf{r}(\alpha\Delta\mathbf{x})\cdot\Delta\mathbf{x}\,d\alpha\right| \le \|\mathbf{r}(\alpha_{\max}\Delta\mathbf{x})\|_2 \|\Delta\mathbf{x}\|_2 \int_0^1 d\alpha = \|\mathbf{r}(\alpha_{\max}\Delta\mathbf{x})\|_2 \|\Delta\mathbf{x}\|_2$$

The following limit is computed to be zero because $\|\mathbf{r}(\alpha_{\max}\Delta\mathbf{x})\|_2$ is $o(\|\Delta\mathbf{x}\|_2)$ as $\Delta\mathbf{x}\to\mathbf{0}$.

$$\lim_{\Delta\mathbf{x}\to\mathbf{0}} \frac{\|\mathbf{r}(\alpha_{\max}\Delta\mathbf{x})\|_2 \|\Delta\mathbf{x}\|_2}{\|\Delta\mathbf{x}\|_2^2} = \lim_{\Delta\mathbf{x}\to\mathbf{0}} \frac{\|\mathbf{r}(\alpha_{\max}\Delta\mathbf{x})\|_2}{\|\Delta\mathbf{x}\|_2} = 0$$

3.4. TAYLOR SERIES

The error term
$$r(\Delta \mathbf{x}) = \int_0^1 \mathbf{r}(\alpha \Delta \mathbf{x}) \cdot \Delta \mathbf{x}\, d\alpha$$
is $o(\|\Delta \mathbf{x}\|_2^2)$ as $\Delta \mathbf{x} \to \mathbf{0}$. □

The following lemma is useful in proving the formulas for 0th order Taylor series.

Lemma 3.4.1. *Suppose $f \in C^1(\Omega)$ where Ω is an open convex subset of $\mathbb{R}^n$ which contains $\mathbf{x}_0$. For every $\mathbf{x} \in \Omega$ we can write $\mathbf{x} = \mathbf{x}_0 + \Delta \mathbf{x}$. If $\phi(\alpha) = f(\mathbf{x}_0 + \alpha \Delta \mathbf{x})$ where $\alpha \in [0,1]$, then $\phi'(\alpha)$ is computed using*

$$\phi'(\alpha) = \nabla f(\mathbf{x}_0 + \alpha \Delta \mathbf{x}) \cdot \Delta \mathbf{x} \qquad (3.4.2)$$

Proof. We use the definition of derivative from equation (3.2.12) to expand about the point $\mathbf{x}_0 + \alpha \Delta \mathbf{x}$.

$$\begin{aligned} f(\mathbf{x}_0) &= f(\mathbf{x}_0 + \alpha \Delta \mathbf{x}) + \nabla f(\mathbf{x}_0 + \alpha \Delta \mathbf{x}) \cdot (-\alpha \Delta \mathbf{x}) + r(-\alpha \Delta \mathbf{x}) \\ f(\mathbf{x}_0 + \alpha \Delta \mathbf{x}) &= f(\mathbf{x}_0) + \nabla f(\mathbf{x}_0 + \alpha \Delta \mathbf{x}) \cdot \alpha \Delta \mathbf{x} - r(-\alpha \Delta \mathbf{x}) \end{aligned}$$

where $r(-\alpha \Delta \mathbf{x})$ is $o(\alpha \|\Delta \mathbf{x}\|_2)$ as $\alpha \Delta \mathbf{x} \to \mathbf{0}$. This equation can also be written as

$$\phi(\alpha) = \phi(0) + [\nabla f(\mathbf{x}_0 + \alpha \Delta \mathbf{x}) \cdot \Delta \mathbf{x}]\alpha + r_\phi(\alpha)$$

where $r_\phi(\alpha)$ is $o(\alpha)$ as $\alpha \to 0$. This is seen if we fix $\Delta \mathbf{x}$ and vary α. Clearly $\alpha \Delta \mathbf{x} \to \mathbf{0}$ as $\alpha \to 0$. Then $r_\phi(\alpha) = -r(-\alpha \Delta \mathbf{x})$ is $o(\alpha \|\Delta \mathbf{x}\|_2) = o(\alpha)$ because $\|\Delta \mathbf{x}\|_2$ is constant. We observe that $\phi'(\alpha) = [\nabla f(\mathbf{x}_0 + \alpha \Delta \mathbf{x}) \cdot \Delta \mathbf{x}]$ by the definition of 1st derivative of a single variable function. □

Theorem 3.4.2 (0th Order Taylor Series). *If $f \in C^1(\Omega)$ where Ω is an open convex subset of $\mathbb{R}^n$ which contains $\mathbf{x}_0$, then for every $\mathbf{x} \in \Omega$ we can write $\mathbf{x} = \mathbf{x}_0 + \Delta \mathbf{x}$ and*

$$\begin{aligned} f(\mathbf{x}_0 + \Delta \mathbf{x}) &= f(\mathbf{x}_0) + \int_0^1 \nabla f(\mathbf{x}_0 + \alpha \Delta \mathbf{x}) \cdot \Delta \mathbf{x}\, d\alpha & (3.4.3) \\ f(\mathbf{x}_0 + \Delta \mathbf{x}) &= f(\mathbf{x}_0) + \nabla f(\mathbf{x}_0 + \mu \Delta \mathbf{x}) \cdot \Delta \mathbf{x} & (3.4.4) \end{aligned}$$

for some $\mu \in [0,1]$.

Proof. The theorem is a simple application of equation (3.3.1) and was used in the proof on the previous page. An alternate proof is given by using equation (3.4.2) in equation (1.4.2).

$$\begin{aligned} \phi(1) - \phi(0) &= \int_0^1 \phi'(\alpha)\, d\alpha \\ f(\mathbf{x}_0 + \Delta \mathbf{x}) - f(\mathbf{x}_0) &= \int_0^1 \nabla f(\mathbf{x}_0 + \alpha \Delta \mathbf{x}) \cdot \Delta \mathbf{x}\, d\alpha \end{aligned}$$

The Integral Mean Value Theorem implies there exists $\mu \in [0,1]$ such that

$$\phi'(\mu) = \int_0^1 \phi'(\alpha)\,d\alpha$$

$$\nabla f(\mathbf{x}_0 + \mu \Delta \mathbf{x}) \cdot \Delta \mathbf{x} = \int_0^1 \nabla f(\mathbf{x}_0 + \alpha \Delta \mathbf{x}) \cdot \Delta \mathbf{x}\, d\alpha$$

$\square$

The 0th order Taylor series is sometimes called the Mean Value Theorem of Vector Calculus in many texts.

Lemma 3.4.2. *Suppose $f \in \mathcal{C}^2(\Omega)$ where Ω is an open convex subset of $\mathbb{R}^n$ which contains $\mathbf{x}_0$. For every $\mathbf{x} \in \Omega$ we can write $\mathbf{x} = \mathbf{x}_0 + \Delta \mathbf{x}$. If $\phi(\alpha) = f(\mathbf{x}_0 + \alpha \Delta \mathbf{x})$ where $\alpha \in [0,1]$, then $\phi''(\alpha)$ is computed using*

$$\phi''(\alpha) = (\Delta \mathbf{x})^T \mathbf{H}(\mathbf{x}_0 + \alpha \Delta \mathbf{x})\, \Delta \mathbf{x} \qquad (3.4.5)$$

Proof. The definition of 2nd derivative from equation (3.2.24) expanded about the point $\mathbf{x}_0 + \alpha \Delta \mathbf{x}$ is

$$\nabla f(\mathbf{x}_0) \cdot \Delta \mathbf{x} = \nabla f(\mathbf{x}_0 + \alpha \Delta \mathbf{x}) \cdot \Delta \mathbf{x} + (-\alpha \Delta \mathbf{x})^T \mathbf{H}(\mathbf{x}_0 + \alpha \Delta \mathbf{x})\, \Delta \mathbf{x}$$
$$+ r(-\alpha \Delta \mathbf{x}, \Delta \mathbf{x})$$
$$\nabla f(\mathbf{x}_0 + \alpha \Delta \mathbf{x}) \cdot \Delta \mathbf{x} = \nabla f(\mathbf{x}_0) \cdot \Delta \mathbf{x} + (\alpha \Delta \mathbf{x})^T \mathbf{H}(\mathbf{x}_0 + \alpha \Delta \mathbf{x})\, \Delta \mathbf{x}$$
$$- r(-\alpha \Delta \mathbf{x}, \Delta \mathbf{x})$$

where $r(-\alpha \Delta \mathbf{x}, \Delta \mathbf{x})$ is $o(\alpha \, \|\Delta \mathbf{x}\|_2^2)$ as $\alpha \Delta \mathbf{x} \to \mathbf{0}$. Using formulas for ϕ and ϕ', we write

$$\phi'(\alpha) = \phi'(0) + [(\Delta \mathbf{x})^T \mathbf{H}(\mathbf{x}_0 + \alpha \Delta \mathbf{x})\, \Delta \mathbf{x}]\alpha + r_\phi(\alpha, \Delta \mathbf{x})$$

where $r_\phi(\alpha, \Delta \mathbf{x})$ is $o(\alpha)$ as $\alpha \to 0$. We show this by fixing $\Delta \mathbf{x}$ and allowing α to vary. Recall that $\alpha \Delta \mathbf{x} \to \mathbf{0}$ as $\alpha \to 0$. Then $r_\phi(\alpha) = r(-\alpha \Delta \mathbf{x}, \Delta \mathbf{x})$ is $o(\alpha \, \|\Delta \mathbf{x}\|_2^2) = o(\alpha)$ because $\|\Delta \mathbf{x}\|_2^2$ is constant. The definition of 2nd derivative of a single variable function implies that $\phi''(\alpha) = (\Delta \mathbf{x})^T \mathbf{H}(\mathbf{x}_0 + \alpha \Delta \mathbf{x})\, \Delta \mathbf{x}$. $\square$

3.4. TAYLOR SERIES

Theorem 3.4.3 (1st Order Taylor Series). *If $f \in C^2(\Omega)$ where Ω is an open convex subset of $\mathbb{R}^n$ which contains $\mathbf{x}_0$, then for every $\mathbf{x} \in \Omega$ we can write $\mathbf{x} = \mathbf{x}_0 + \Delta\mathbf{x}$ and*

$$\begin{aligned} f(\mathbf{x}_0 + \Delta\mathbf{x}) &= f(\mathbf{x}_0) + \nabla f(\mathbf{x}_0) \cdot \Delta\mathbf{x} \\ &+ \int_0^1 (1-\alpha)(\Delta\mathbf{x})^T \mathbf{H}(\mathbf{x}_0 + \alpha\Delta\mathbf{x})\, \Delta\mathbf{x}\, d\alpha \end{aligned} \quad (3.4.6)$$

$$\begin{aligned} f(\mathbf{x}_0 + \Delta\mathbf{x}) &= f(\mathbf{x}_0) + \nabla f(\mathbf{x}_0) \cdot \Delta\mathbf{x} \\ &+ \frac{1}{2}(\Delta\mathbf{x})^T \mathbf{H}(\mathbf{x}_0 + \mu\Delta\mathbf{x})\, \Delta\mathbf{x} \end{aligned} \quad (3.4.7)$$

for some $\mu \in [0,1]$.

Proof. Set $\phi(\alpha) = f(\mathbf{x}_0 + \alpha\Delta\mathbf{x})$ and use equations (3.4.2) and (3.4.5) in equation (1.4.3).

$$\begin{aligned} \phi(1) &= \phi(0) + \phi'(0) + \int_0^1 (1-\alpha)\phi''(\alpha)\, d\alpha \\ f(\mathbf{x}_0 + \Delta\mathbf{x}) &= f(\mathbf{x}_0) + \nabla f(\mathbf{x}_0) \cdot \Delta\mathbf{x} \\ &+ \int_0^1 (1-\alpha)(\Delta\mathbf{x})^T \mathbf{H}(\mathbf{x}_0 + \alpha\Delta\mathbf{x})\, \Delta\mathbf{x}\, d\alpha \end{aligned}$$

The General Integral Mean Value Theorem is applied to the integral remainder term. There exists $\mu \in [0,1]$ such that

$$\begin{aligned} \int_0^1 (1-\alpha)(\Delta\mathbf{x})^T \mathbf{H}(\mathbf{x}_0 + \alpha\Delta\mathbf{x})\, \Delta\mathbf{x}\, d\alpha &= (\Delta\mathbf{x})^T \mathbf{H}(\mathbf{x}_0 + \mu\Delta\mathbf{x})\, \Delta\mathbf{x} \int_0^1 (1-\alpha)\, d\alpha \\ &= \frac{1}{2}(\Delta\mathbf{x})^T \mathbf{H}(\mathbf{x}_0 + \mu\Delta\mathbf{x})\, \Delta\mathbf{x} \end{aligned}$$

$\square$

3.5 Local Minimums

The major definitions and theorems for local minimums of multivariable functions are provided in this section. The domain of the functions will be an open convex set Ω. The proofs of the theorems are similar to the ones for single variable functions.

Definition 3.5.1 (Stationary Point). $\mathbf{x}^*$ is the location of a *stationary point* of the function $f(\mathbf{x})$ if $\nabla f(\mathbf{x}^*) = \mathbf{0}$.

For differentiable functions, being a stationary point is a necessary condition for a minimum.

Definition 3.5.2 (Local Minimum). $\mathbf{x}^*$ is the location of a *local minimum* of $f(\mathbf{x})$ if there exists $\delta > 0$ such that $f(\mathbf{x}^*) \leq f(\mathbf{x})$ for all $\mathbf{x} \in B(\mathbf{x}^*, \delta)$.

Theorem 3.5.1 (1st Order Necessary Condition). *If $\mathbf{x}^*$ is the location of a local minimum of $f(\mathbf{x})$ and $f \in \mathcal{C}^1(\Omega)$, then $\nabla f(\mathbf{x}^*) = \mathbf{0}$.*

Proof. The proof is by contradiction. Suppose that $\nabla f(\mathbf{x}^*) \neq \mathbf{0}$. Set $\Delta \mathbf{x} = -\alpha \nabla f(\mathbf{x}^*)$ where $\alpha > 0$ is small enough so that $\mathbf{x}^* + \Delta \mathbf{x} \in \Omega$. According to the definition of 1st derivative, we have

$$f(\mathbf{x}^* - \alpha \nabla f(\mathbf{x}^*)) = f(\mathbf{x}^*) - \alpha \|\nabla f(\mathbf{x}^*)\|_2^2 + r(-\alpha \nabla f(\mathbf{x}^*))$$

The error term is $o(\alpha \|\nabla f(\mathbf{x}^*)\|_2)$ and we can find $\delta > 0$ such that $|r(-\alpha \nabla f(\mathbf{x}^*))| < \frac{1}{2}\alpha \|\nabla f(\mathbf{x}^*)\|_2^2$ when $0 < \alpha \|\nabla f(\mathbf{x}^*)\|_2 < \delta$. This procedure sets a limit on the size of α and we set $\bar{\alpha} = \delta / \|\nabla f(\mathbf{x}^*)\|_2$. Under these conditions

$$f(\mathbf{x}^* - \alpha \nabla f(\mathbf{x}^*)) \leq f(\mathbf{x}^*) - \frac{1}{2}\alpha \|\nabla f(\mathbf{x}^*)\|_2^2 < f(\mathbf{x}^*)$$

for all $\alpha \in (0, \bar{\alpha})$. This contradicts the assumption that x^* is a local minimum. Thus $\nabla f(\mathbf{x}^*) = \mathbf{0}$. $\square$

The 2nd order conditions require the notion of positive semi-definite and positive definite bilinear forms.

Definition 3.5.3 (Positive Semi-definite). The bilinear form $D^2(\mathbf{x})[\mathbf{u}, \mathbf{v}]$ is *positive semi-definite* if $D^2[\mathbf{u}, \mathbf{u}] \geq 0$ for all $\mathbf{u} \in \mathbb{R}^n$ such that $\mathbf{u} \neq \mathbf{0}$.

Definition 3.5.4 (Positive Definite). The bilinear form $D^2(\mathbf{x})[\mathbf{u}, \mathbf{v}]$ is *positive definite* if $D^2[\mathbf{u}, \mathbf{u}] > 0$ for all $\mathbf{u} \in \mathbb{R}^n$ such that $\mathbf{u} \neq \mathbf{0}$.

3.5. LOCAL MINIMUMS

Theorem 3.5.2 (2nd Order Necessary Condition). *If $\mathbf{x}^*$ is the location of a local minimum of $f(\mathbf{x})$ and $f \in \mathcal{C}^2(\Omega)$, then $\nabla f(\mathbf{x}^*) = \mathbf{0}$ and $\mathbf{H}(\mathbf{x}^*)$ is positive semi-definite.*

Proof. The 1st order necessary condition implies $\nabla f(\mathbf{x}^*) = \mathbf{0}$. The remainder of the proof is by contradiction. Suppose there is a direction $\Delta \mathbf{x}$ so that

$$(\Delta \mathbf{x})^T \mathbf{H}(\mathbf{x}^*) \Delta \mathbf{x} < 0 \quad \text{and} \quad \mathbf{x}^* + \Delta \mathbf{x} \in \Omega$$

Then the definition of 2nd derivative with $\nabla f(\mathbf{x}^*) = \mathbf{0}$ gives

$$f(\mathbf{x}^* + \alpha \Delta \mathbf{x}) = f(\mathbf{x}^*) + \frac{1}{2}\alpha^2 (\Delta \mathbf{x})^T \mathbf{H}(\mathbf{x}^*) \Delta \mathbf{x} + r(\alpha \Delta \mathbf{x})$$

for all $\alpha \in [0, 1]$. The error term is $o(\alpha^2 \|\mathbf{x}\|_2^2)$ and we can find $\delta > 0$ such that $|r(\alpha \Delta \mathbf{x})| < \frac{1}{4}\alpha^2 |(\Delta \mathbf{x})^T \mathbf{H}(\mathbf{x}^*) \Delta \mathbf{x}|$ when $0 < \alpha \|\Delta \mathbf{x}\|_2 < \delta$. Set $\bar{\alpha} = \delta / \|\Delta \mathbf{x}\|_2$ and observe that

$$f(\mathbf{x}^* + \alpha \Delta \mathbf{x}) \leq f(\mathbf{x}^*) + \frac{1}{4}\alpha^2 (\Delta \mathbf{x})^T \mathbf{H}(\mathbf{x}^*) \Delta \mathbf{x} < f(\mathbf{x}^*)$$

for all $\alpha \in (0, \bar{\alpha})$. This contradicts the assumption that $\mathbf{x}^*$ is a local minimum. Thus $\mathbf{H}(\mathbf{x}^*)$ must be positive semi-definite. □

For twice continuously differentiable functions, the next theorem provides a sufficient condition for a local minimum.

Theorem 3.5.3 (2nd Order Sufficient Condition). *If $f \in \mathcal{C}^2(\Omega)$, $\nabla f(\mathbf{x}^*) = \mathbf{0}$, and $\mathbf{H}(\mathbf{x}^*)$ is positive definite, then $\mathbf{x}^*$ is the location of a local minimum of $f(\mathbf{x})$.*

Proof. Since $f \in \mathcal{C}^2(\Omega)$, there exists $\delta > 0$ such that $\mathbf{H}(\mathbf{x})$ is positive definite for all $\mathbf{x} \in B(\mathbf{x}^*, \delta) \subset \Omega$. The 1st order Taylor series expansion of $f(\mathbf{x})$ about $\mathbf{x}^*$ is

$$f(\mathbf{x}^* + \Delta \mathbf{x}) = f(\mathbf{x}^*) + \nabla f(\mathbf{x}^*) \cdot \Delta \mathbf{x} + \frac{1}{2}(\Delta \mathbf{x})^T \mathbf{H}(\mathbf{x}^* + \mu \Delta \mathbf{x}) \Delta \mathbf{x}$$

where $\mathbf{x} = \mathbf{x}^* + \Delta \mathbf{x} \in B(\mathbf{x}^*, \delta)$ and $\mu \in [0, 1]$. Since $\nabla f(\mathbf{x}^*) = \mathbf{0}$ and $\mathbf{H}(\mathbf{x}^*)$ is positive definite, we clearly have $f(\mathbf{x}) \geq f(\mathbf{x}^*)$ for all $\mathbf{x} \in B(\mathbf{x}^*, \delta)$. Thus $\mathbf{x}^*$ is the location of a local minimum of $f(\mathbf{x})$. □

3.6 Variation Notation

The variation $\delta \mathbf{x}$ of the independent variable $\mathbf{x}$ is any arbitrary increment in $\mathbf{x}$. We set $\delta \mathbf{x} = \Delta \mathbf{x}$ and formulate definitions and theorems concerning local minimums in variation notation. Let Ω be an open convex subset of $\mathbb{R}^n$.

Definition 3.6.1 (Admissible). The variation $\delta \mathbf{x}$ at $\mathbf{x} \in \Omega$ is *admissible* if $\mathbf{x} + \delta \mathbf{x} \in \Omega$.

Definition 3.6.2 (1st Variation). The *1st variation* of a function $f \in \mathcal{C}^1(\Omega)$ is defined by $\delta f \equiv \nabla f(\mathbf{x}) \cdot \delta \mathbf{x}$ where $\delta \mathbf{x}$ is any admissible variation at $\mathbf{x} \in \Omega$.

Definition 3.6.3 (2nd Variation). The *2nd variation* of a function $f \in \mathcal{C}^2(\Omega)$ is defined by $\delta^2 f \equiv \frac{1}{2}(\delta \mathbf{x})^T \mathbf{H}(\mathbf{x}) \, \delta \mathbf{x}$ where $\delta \mathbf{x}$ is any admissible variation at $\mathbf{x} \in \Omega$.

Definition 3.6.4 (Stationary Point). $\mathbf{x}^*$ is the location of a *stationary point* of the function $f(\mathbf{x})$ if $\delta f = 0$ for all admissible variations $\delta \mathbf{x}$ at $\mathbf{x}^*$.

Theorem 3.6.1 (1st Order Necessary Condition). *If $\mathbf{x}^*$ is the location of a local minimum of $f(\mathbf{x})$ and $f \in \mathcal{C}^1(\Omega)$, then $\delta f = 0$ for all admissible variations $\delta \mathbf{x}$ at $\mathbf{x}^*$.*

Theorem 3.6.2 (2nd Order Necessary Condition). *If $\mathbf{x}^*$ is the location of a local minimum of $f(\mathbf{x})$ and $f \in \mathcal{C}^2(\Omega)$, then $\delta f = 0$ and $\delta^2 f \geq 0$ for all admissible variations $\delta \mathbf{x}$ at $\mathbf{x}^*$.*

Theorem 3.6.3 (2nd Order Sufficient Condition). *If $f \in \mathcal{C}^2(\Omega)$, $\delta f = 0$, and $\delta^2 f > 0$ for all admissible variations δx at $\mathbf{x}^*$, then $\mathbf{x}^*$ is the location of a local minimum of $f(\mathbf{x})$.*

3.7 Constrained Minimization

It is common to require minimization of a function with constraints in many applications. Let Ω be an open convex subset of $\mathbb{R}^n$ and $f(\mathbf{x}) \in \mathcal{C}^1(\Omega)$. The constraints are given by m equations: $g_k(\mathbf{x}) = 0$. We require $m < n$ and $g_k \in \mathcal{C}^1(\Omega)$ for $k = 1, \ldots, m$. Variation notation is used in this section and we write the resulting equations in component form.

The minimization problem to be solved is

$$\mathbf{x}^* = \arg \operatorname*{local\,min}_{\mathbf{x} \in \Omega} f(\mathbf{x}) \text{ subject to } g_k(\mathbf{x}) = 0, \quad k = 1, \ldots, m \quad (3.7.1)$$

$\mathbf{x}^*$ is the location of a local minimum. We notice that the constraints imply that not all the components of $\mathbf{x}$ are independent. There are only $n - m$ independent variables. The remaining m variables depend on the independent variables. Attempting to reformulate the problem with fewer variables is impractical since the constraint equations are usually nonlinear. Instead, we use the method of Lagrange multipliers which produces necessary conditions for a local minimum.

The 1st order necessary condition requires $\delta f = \nabla f(\mathbf{x}) \cdot \delta \mathbf{x} = 0$. In component form, the equations are

$$\sum_{i=1}^{n} \frac{\partial}{\partial x_i} f(\mathbf{x}) \, \delta x_i = 0 \quad (3.7.2)$$

If δx_i could be arbitrarily varied, then each $\frac{\partial}{\partial x_i} f(\mathbf{x}) = 0$. However, the components of $\delta \mathbf{x}$ are not independent and cannot be arbitrarily varied. The method of Lagrange multipliers to solve (3.7.1) will now be formulated.

First, we compute the 1st variation of each constraint equation. It is obvious that if a function is equal to zero, then its variation is also equal to zero.

$$\delta g_k = \nabla g_k(\mathbf{x}) \cdot \delta \mathbf{x} = \sum_{i=1}^{n} \frac{\partial}{\partial x_i} g_k(\mathbf{x}) \, \delta x_i = 0, \quad k = 1, \ldots, m \quad (3.7.3)$$

We introduce the variables $\lambda_1, \ldots, \lambda_m$ which will allow us to perform the minimization. These variables are known as the *Lagrange multipliers* and their values are determined while solving the minimization problem. Each equation in (3.7.3) is multiplied by a corresponding Lagrange multiplier.

$$\lambda_k \sum_{i=1}^{n} \frac{\partial}{\partial x_i} g_k(\mathbf{x}) \, \delta x_i = 0, \quad k = 1, \ldots, m \quad (3.7.4)$$

We add the sum of the m equations in (3.7.4) to equation (3.7.2) to get

$$\sum_{i=1}^{n} \frac{\partial}{\partial x_i} f(\mathbf{x}) \, \delta x_i + \sum_{k=1}^{m} \left[\lambda_k \sum_{i=1}^{n} \frac{\partial}{\partial x_i} g_k(\mathbf{x}) \, \delta x_i \right] = 0$$

Rearranging the sum and factoring out the variations δx_i produces

$$\sum_{i=1}^{n} \left[\frac{\partial}{\partial x_i} f(\mathbf{x}) + \sum_{k=1}^{m} \lambda_k \frac{\partial}{\partial x_i} g_k(\mathbf{x}) \right] \delta x_i = 0 \tag{3.7.5}$$

It must be emphasized that the variations δx_i are not independent. Thus, we cannot immediately conclude that the term in brackets vanishes.

Consider the first m variables $x_1, \ldots, x_m$ to be dependent and the remaining $x_{m+1}, \ldots, x_n$ as independent. We partition the sum in equation (3.7.5).

$$\sum_{i=1}^{m} \left[\frac{\partial}{\partial x_i} f(\mathbf{x}) + \sum_{k=1}^{m} \lambda_k \frac{\partial}{\partial x_i} g_k(\mathbf{x}) \right] \delta x_i$$
$$+ \sum_{i=m+1}^{n} \left[\frac{\partial}{\partial x_i} f(\mathbf{x}) + \sum_{k=1}^{m} \lambda_k \frac{\partial}{\partial x_i} g_k(\mathbf{x}) \right] \delta x_i = 0 \tag{3.7.6}$$

We choose values of λ_k in the first sum which causes the term in every bracket to vanish.

$$\frac{\partial}{\partial x_i} f(\mathbf{x}) + \sum_{k=1}^{m} \lambda_k \frac{\partial}{\partial x_i} g_k(\mathbf{x}) = 0, \qquad i = 1, \ldots, m$$

The values for λ_k can be determined by solving the matrix system

$$\begin{bmatrix} \frac{\partial}{\partial x_1} g_1(\mathbf{x}) & \cdots & \frac{\partial}{\partial x_1} g_m(\mathbf{x}) \\ \vdots & \ddots & \vdots \\ \frac{\partial}{\partial x_m} g_1(\mathbf{x}) & \cdots & \frac{\partial}{\partial x_m} g_m(\mathbf{x}) \end{bmatrix} \begin{bmatrix} \lambda_1 \\ \vdots \\ \lambda_m \end{bmatrix} = \begin{bmatrix} -\frac{\partial}{\partial x_1} f(\mathbf{x}) \\ \vdots \\ -\frac{\partial}{\partial x_m} f(\mathbf{x}) \end{bmatrix}$$

The solution exists if the determinant of the matrix is nonzero.

The second sum in equation (3.7.6) involves variables which can be independently varied. Each term in brackets for that sum is required to vanish.

$$\frac{\partial}{\partial x_i} f(\mathbf{x}) + \sum_{k=1}^{m} \lambda_k \frac{\partial}{\partial x_i} g_k(\mathbf{x}) = 0, \qquad i = m+1, \ldots, n$$

We now have that every term in brackets vanishes in equation (3.7.6). This completes the derivation and the method is summarized in the next theorem.

3.7. CONSTRAINED MINIMIZATION

Theorem 3.7.1 (1st Order Necessary Condition with Constraints). *Let Ω be an open convex subset of $\mathbb{R}^n$. Suppose $f \in \mathcal{C}^1(\Omega)$ and $g_k \in \mathcal{C}^1(\Omega)$ for $k = 1, \ldots, m$ where $m < n$. If $\mathbf{x}^*$ is the location of a local minimum of $f(\mathbf{x})$ subject to the constraints $g_k(\mathbf{x}) = 0$, then $\mathbf{x}^*$ is a solution to*

$$\frac{\partial}{\partial x_i} f(\mathbf{x}) + \sum_{k=1}^{m} \lambda_k \frac{\partial}{\partial x_i} g_k(\mathbf{x}) = 0, \qquad i = 1, \ldots, n \qquad (3.7.7)$$

$$g_k(\mathbf{x}) = 0, \qquad k = 1, \ldots, m \qquad (3.7.8)$$

for some constants $\lambda_1, \ldots, \lambda_m$.

Equations (3.7.7) and (3.7.8) are $n + m$ equations for the $n + m$ variables: $x_1, \ldots, x_n, \lambda_1, \ldots, \lambda_m$. Equivalently, we could formulate the method as the unconstrained minimization of the augmented function

$$F(\mathbf{u}) = f(\mathbf{x}) + \sum_{k=1}^{m} \lambda_k g_k(\mathbf{x}) \qquad (3.7.9)$$

where $\mathbf{u} = [x_1, \ldots, x_n, \lambda_1, \ldots, \lambda_m]^T$. The necessary condition $\delta F(\mathbf{u}) = 0$ results in equations (3.7.7) and (3.7.8).

Part II

Calculus of Variations for Single Functions

Chapter 4

Formulation

4.1 Introduction

The goal of calculus of variations is to find a function $y \in \mathcal{U}$ which minimizes a functional $J : \mathcal{U} \to \mathbb{R}$. The set $\mathcal{U}$ consists of admissible functions $y : [a, b] \to \mathbb{R}$ which satisfy boundary conditions at the endpoints. The exact specification of $\mathcal{U}$ is formulated later but it will be a subset of $\mathcal{C}^n([a, b])$. The order n depends on the problem being solved. It must be emphasized that the set $\mathcal{U}$ might not be a vector space.

An example of a functional is

$$J[y] = \int_a^b F(x, y, y', y'') \, dx \qquad (4.1.1)$$

The function F is assumed to have continuous partial derivatives of its arguments up to order 2.

Notice that a functional maps functions to real numbers. In contrast to the problems in Part I of this book, we seek a solution which is a function instead of a point. We state this problem as

$$y^* = \arg \operatorname*{local\,min}_{y \in \mathcal{U}} J[y] \qquad (4.1.2)$$

where y^* is the location of a local minimum of J.

Problem (4.1.2) is solved by computing the variation of the functional J. The variation of a functional, with a function as its argument, is more complex than for ordinary functions. This explains why many comprehensive texts have been written on the calculus of variations [7, 9, 21].

4.2 General Functionals

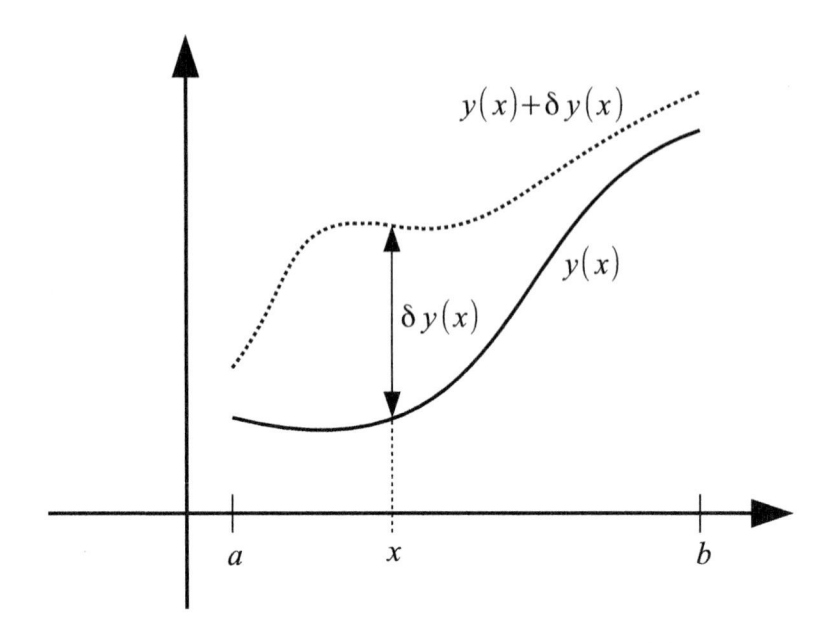

Figure 4.2.1: Variation $\delta y(x)$ of function $y(x)$

In order to formulate the variation of the functional J, we must first define the variation of the independent variable y. Since y is a function of x, its variation δy will also be a function of x. A simple way to understand the variation δy is provided in figure 4.2.1. For a fixed value of x, the variation is the difference between the values of functions $y(x) + \delta y(x)$ and $y(x)$.

The variation δy is *admissible* if $y + \delta y \in \mathcal{U}$. This requirement restricts the choice of δy. Let $\mathcal{V}$ denote the set of admissible variations. $\mathcal{U} \subset \mathcal{C}^n([a,b])$ implies that $\mathcal{V} \subset \mathcal{C}^n([a,b])$. We also require that $\mathcal{V}$ be a subspace of $\mathcal{C}^n([a,b])$.

Definition 4.2.1 (1st Variation). The functional $J[y]$ has a *1st variation* if there exists a linear functional $\delta J[y] : \mathcal{V} \to \mathbb{R}$ such that

$$J[y + \delta y] = J[y] + \delta J[y][\delta y] + \varepsilon \, \|\delta y\|_{nsup} \qquad (4.2.1)$$

where $y + \delta y \in \mathcal{U}$ and $\varepsilon \to 0$ as $\|\delta y\|_{nsup} \to 0$.

Note that $\delta y \to 0$ means $\delta y(x) \to 0$ for all $x \in [a,b]$. $\delta J[y][\delta y]$ depends on the function y and is a linear functional on its variation δy. The order of the $nsup$ norm depends on the set of admissible variations $\mathcal{V}$.

4.2. GENERAL FUNCTIONALS

Some additional properties of the 1st variation will be useful in later sections and chapters. We first define scalar multiplication and vector addition for functionals. Let $\alpha \in \mathbb{R}$ and J and K be functionals on $\mathcal{U}$.

$$(\alpha J)[y] = \alpha J[y]$$
$$(J + K)[y] = J[y] + K[y]$$

We show that the 1st variation is a linear operator.

Lemma 4.2.1. *The 1st variation is a linear operator. If J and K are functionals on $\mathcal{U}$ with a 1st variation and $\alpha \in \mathbb{R}$, then for all $y \in \mathcal{U}$ and admissible variations $\delta y \in \mathcal{V}$*

$$\delta(\alpha J)[y][\delta y] = \alpha \, \delta J[y][\delta y]$$
$$\delta(J + K)[y][\delta y] = \delta J[y][\delta y] + \delta K[y][\delta y]$$

Proof. Using the fact that $(\alpha J)[y] = \alpha J[y]$ and the definition of 1st variation of J, we write

$$\begin{aligned}(\alpha J)[y + \delta y] &= \alpha J[y + \delta y] \\ &= \alpha \left\{ J[y] + \delta J[y][\delta y] + \varepsilon \left\| \delta y \right\|_{nsup} \right\} \\ &= \alpha J[y] + \alpha \, \delta J[y][\delta y] + \alpha \varepsilon \left\| \delta y \right\|_{nsup} \end{aligned}$$

where $\varepsilon \to 0$ as $\left\| \delta y \right\|_{nsup} \to 0$. Now define $\tilde{\varepsilon} = \alpha \varepsilon$ and write

$$(\alpha J)[y + \delta y] = (\alpha J)[y] + \alpha \, \delta J[y][\delta y] + \tilde{\varepsilon} \left\| \delta y \right\|_{nsup}$$

We have $\tilde{\varepsilon} \to 0$ as $\left\| \delta y \right\|_{nsup} \to 0$. The 1st variation must be defined as $\delta(\alpha J)[y][\delta y] = \alpha \, \delta J[y][\delta y]$.

The linearity property of addition is proved by using $(J+K)[y] = J[y]+K[y]$.

$$\begin{aligned}(J+K)[y+\delta y] &= J[y+\delta y] + K[y+\delta y] \\ &= J[y] + \delta J[y][\delta y] + \varepsilon_1 \left\| \delta y \right\|_{nsup} \\ &\quad + K[y] + \delta K[y][\delta y] + \varepsilon_2 \left\| \delta y \right\|_{nsup} \\ &= J[y] + K[y] + \delta J[y][\delta y] + \delta K[y][\delta y] \\ &\quad + (\varepsilon_1 + \varepsilon_2) \left\| \delta y \right\|_{nsup}\end{aligned}$$

where $\varepsilon_1, \varepsilon_2 \to 0$ as $\left\| \delta y \right\|_{nsup} \to 0$. Let $\tilde{\varepsilon} = \varepsilon_1 + \varepsilon_2$ and we observe that

$$(J+K)[y+\delta y] = (J+K)[y] + \delta J[y][\delta y] + \delta K[y][\delta y] + \tilde{\varepsilon} \left\| \delta y \right\|_{nsup}$$

Hence $\delta(J+K)[y][\delta y] = \delta J[y][\delta y] + \delta K[y][\delta y]$ because $\tilde{\varepsilon} \to 0$ as $\left\| \delta y \right\|_{nsup} \to 0$. □

Proving the uniqueness of the 1st variation requires the application of the following lemma.

Lemma 4.2.2. *Let $\mathcal{V}$ be a subspace of $\mathcal{C}^n([a,b])$. If $\varphi : \mathcal{V} \to \mathbb{R}$ is a linear functional such that $\varphi[v]/\|v\|_{nsup} \to 0$ as $\|v\|_{nsup} \to 0$, then $\varphi[v] = 0$ for all $v \in \mathcal{V}$.*

Proof. The proof is by contradiction. Suppose that $\varphi[v] \neq 0$. Choose $v_0 \in \mathcal{V}$ such that $v_0 \neq 0$ and $\varphi[v_0] = \varphi_0 \neq 0$. Clearly $\|v_0\|_{nsup} \neq 0$. Let $v_k = v_0/k$. Then $v_k \to 0$ as $k \to \infty$. Since $v_k \in \mathcal{V}$, we also have $\|v_k\|_{nsup} \to 0$ as $k \to \infty$.

$$\lim_{k \to \infty} \frac{\varphi[v_k]}{\|v_k\|_{nsup}} = \lim_{k \to \infty} \frac{\varphi[v_0/k]}{\|v_0/k\|_{nsup}} = \lim_{k \to \infty} \frac{\frac{1}{k}\varphi[v_0]}{\frac{1}{k}\|v_0\|_{nsup}} = \frac{\varphi_0}{\|v_0\|_{nsup}} \neq 0$$

This contradicts the hypothesis of the theorem. Therefore, we must have $\varphi[v] = 0$ for all $v \in \mathcal{V}$. $\square$

Lemma 4.2.3. *The 1st variation of a functional is unique.*

Proof. Let δJ_1 and δJ_2 be 1st variations of the functional J. Then by the definition of 1st variation, we can write

$$\begin{aligned} J[y+\delta y] &= J[y] + \delta J_1[y][\delta y] + \varepsilon_1 \|\delta y\|_{nsup} \\ J[y+\delta y] &= J[y] + \delta J_2[y][\delta y] + \varepsilon_2 \|\delta y\|_{nsup} \end{aligned}$$

where where $y + \delta y \in \mathcal{U}$ and $\varepsilon_1, \varepsilon_2 \to 0$ as $\|\delta y\|_{nsup} \to 0$. Subtracting the two equations above and rearranging

$$\delta J_1[y][\delta y] - \delta J_2[y][\delta y] = (\varepsilon_2 - \varepsilon_1) \|\delta y\|_{nsup}$$

The left hand side is the linear functional $\varphi = (\delta J_1 - \delta J_2)[y]$.

$$\varphi[\delta y] = (\varepsilon_2 - \varepsilon_1) \|\delta y\|_{nsup}$$

Divide both sides by $\|\delta y\|_{nsup}$ and let $\|\delta y\|_{nsup} \to 0$.

$$\frac{\varphi[\delta y]}{\|\delta y\|_{nsup}} \to 0$$

Using the previous lemma, this implies that $\varphi[\delta y] = 0$ and hence $\delta J_1 = \delta J_2$. $\square$

4.2. GENERAL FUNCTIONALS

Instead of defining the 2nd variation as the variation of the 1st variation, we prefer a Taylor series approach.

Definition 4.2.2 (2nd Variation). The functional $J[y]$ has a *2nd variation* if it has a first variation and there exists a quadratic functional $\delta^2 J[y] : \mathcal{V} \to \mathbb{R}$ such that
$$J[y + \delta y] = J[y] + \delta J[y][\delta y] + \delta^2 J[y][\delta y] + \varepsilon \, \|\delta y\|_{nsup}^2 \qquad (4.2.2)$$
where $y + \delta y \in \mathcal{U}$ and $\varepsilon \to 0$ as $\|\delta y\|_{nsup} \to 0$.

We clearly recognize equation (4.2.2) as a 2nd order Taylor series. The factor of $1/2$ is absorbed into $\delta^2 J$ and this will be observed when computing the 2nd variation.

Lemma 4.2.4. *The 2nd variation is a linear operator. If J and K are functionals on $\mathcal{U}$ with a 2nd variation and $\alpha \in \mathbb{R}$, then for all $y \in \mathcal{U}$ and admissible variations $\delta y \in \mathcal{V}$*
$$\begin{aligned} \delta^2(\alpha J)[y][\delta y] &= \alpha \, \delta^2 J[y][\delta y] \\ \delta^2(J + K)[y][\delta y] &= \delta^2 J[y][\delta y] + \delta^2 K[y][\delta y] \end{aligned}$$

Proof. The definition of 2nd variation of J gives
$$\begin{aligned} (\alpha J)[y + \delta y] &= \alpha \, J[y + \delta y] \\ &= \alpha \left\{ J[y] + \delta J[y][\delta y] + \delta^2 J[y][\delta y] + \varepsilon \, \|\delta y\|_{nsup}^2 \right\} \\ &= \alpha \, J[y] + \alpha \, \delta J[y][\delta y] + \alpha \, \delta^2 J[y][\delta y] + \alpha\varepsilon \, \|\delta y\|_{nsup}^2 \end{aligned}$$
where $\varepsilon \to 0$ as $\|\delta y\|_{nsup} \to 0$. Using the fact that the 1st variation is a linear operator and defining $\tilde{\varepsilon} = \alpha\varepsilon$.
$$(\alpha J)[y + \delta y] = (\alpha J)[y] + \delta(\alpha J)[y][\delta y] + \alpha \, \delta^2 J[y][\delta y] + \tilde{\varepsilon} \, \|\delta y\|_{nsup}^2$$
We have $\tilde{\varepsilon} \to 0$ as $\|\delta y\|_{nsup} \to 0$. The 2nd variation must be defined as $\delta^2(\alpha J)[y][\delta y] = \alpha \, \delta^2 J[y][\delta y]$.

The linearity property of addition is proved as follows:
$$\begin{aligned} (J + K)[y + \delta y] &= J[y + \delta y] + K[y + \delta y] \\ &= J[y] + \delta J[y][\delta y] + \delta^2 J[y][\delta y] + \varepsilon_1 \, \|\delta y\|_{nsup}^2 \\ &\quad + K[y] + \delta K[y][\delta y] + \delta^2 K[y][\delta y] + \varepsilon_2 \, \|\delta y\|_{nsup}^2 \\ &= J[y] + K[y] + \delta J[y] + \delta K[y] \\ &\quad + \delta^2 J[y][\delta y] + \delta^2 K[y][\delta y] + (\varepsilon_1 + \varepsilon_2) \, \|\delta y\|_{nsup}^2 \end{aligned}$$

$\varepsilon_1, \varepsilon_2 \to 0$ as $\|\delta y\|_{nsup} \to 0$. Let $\tilde{\varepsilon} = \varepsilon_1 + \varepsilon_2$ and notice that

$$(J+K)[y+\delta y] = (J+K)[y] + \delta(J+K)[y][\delta y] + \delta^2 J[y][\delta y] + \delta^2 K[y][\delta y] + \tilde{\varepsilon}\|\delta y\|_{nsup}$$

Hence $\delta^2(J+K)[y][\delta y] = \delta^2 J[y][\delta y] + \delta^2 K[y][\delta y]$ because $\tilde{\varepsilon} \to 0$ as $\|\delta y\|_{nsup} \to 0$. □

The uniqueness of the 2nd variation is proved in the following lemmas.

Lemma 4.2.5. *Let $\mathcal{V}$ be a subspace of $\mathcal{C}^n([a,b])$. If $\psi : \mathcal{V} \to \mathbb{R}$ is a quadratic functional such that $\psi[v]/\|v\|_{nsup}^2 \to 0$ as $\|v\|_{nsup} \to 0$, then $\psi[v] = 0$ for all $v \in \mathcal{V}$.*

Proof. The proof is by contradiction. Suppose that $\psi[v] \neq 0$. Choose $v_0 \in \mathcal{V}$ such that $v_0 \neq 0$ and $\psi[v_0] = \psi_0 \neq 0$. Clearly $\|v_0\|_{nsup} \neq 0$. Let $v_k = v_0/k$. Then $v_k \to 0$ as $k \to \infty$. Since $v_k \in \mathcal{V}$, we also have $\|v_k\|_{nsup} \to 0$ as $k \to \infty$.

$$\lim_{k\to\infty} \frac{\psi[v_k]}{\|v_k\|_{nsup}^2} = \lim_{k\to\infty} \frac{\psi[v_0/k]}{\|v_0/k\|_{nsup}^2} = \lim_{k\to\infty} \frac{\frac{1}{k^2}\psi[v_0]}{\frac{1}{k^2}\|v_0\|_{nsup}^2} = \frac{\psi_0}{\|v_0\|_{nsup}^2} \neq 0$$

This contradicts the hypothesis of the theorem. Therefore, we must have $\psi[v] = 0$ for all $v \in \mathcal{V}$. □

Lemma 4.2.6. *The 2nd variation of a functional is unique.*

Proof. Let $\delta^2 J_1$ and $\delta^2 J_2$ be 2nd variations of the functional J. Then by the definition of 2nd variation, we can write

$$J[y+\delta y] = J[y] + \delta J[y][\delta y] + \delta^2 J_1[y][\delta y] + \varepsilon_1 \|\delta y\|_{nsup}^2$$
$$J[y+\delta y] = J[y] + \delta J[y][\delta y] + \delta^2 J_2[y][\delta y] + \varepsilon_2 \|\delta y\|_{nsup}^2$$

where where $y + \delta y \in \mathcal{U}$ and $\varepsilon_1, \varepsilon_2 \to 0$ as $\|\delta y\|_{nsup} \to 0$. Subtracting the two equations above and rearranging, we have

$$\delta^2 J_1[y][\delta y] - \delta^2 J_2[y][\delta y] = (\varepsilon_2 - \varepsilon_1) \|\delta y\|_{nsup}^2$$

The left hand side is the quadratic functional $\psi = (\delta^2 J_1 - \delta^2 J_2)[y]$.

$$\psi[\delta y] = (\varepsilon_2 - \varepsilon_1) \|\delta y\|_{nsup}^2$$

Divide both sides by $\|\delta y\|_{nsup}^2$ and let $\|\delta y\|_{nsup} \to 0$.

$$\frac{\psi[\delta y]}{\|\delta y\|_{nsup}^2} \to 0$$

Using the previous lemma, this implies that $\psi[\delta y] = 0$ and hence $\delta^2 J_1 = \delta^2 J_2$. □

4.2. GENERAL FUNCTIONALS

It must be emphasized that δ^2 is a linear operator when applied to functionals. The resulting functional is always a quadratic functional. This is analogous to the differential operator d^2/dx^2. It has the properties of a linear operator but the resulting function can be nonlinear.

In order to define continuity classes of functionals, we recall the definition of a continuous functional on a function space.

Definition 4.2.3 (Continuous Functional). $J[y]$ is a *continuous functional* on $\mathcal{U} \subset \mathcal{C}^n([a,b])$ if for every $\varepsilon > 0$, there exists $\delta > 0$ such that $|J[y] - J[y_0]| < \varepsilon$ whenever $\|y - y_0\|_{nsup} < \delta$ where $y, y_0 \in \mathcal{U}$.

Defining a continuous 1st variation $\delta J[y][\delta y]$ is complicated by the fact that it depends on two arguments y and δy. First, we require that $\delta J[y][\delta y]$ be a continuous linear functional in δy. Next, hold δy fixed and require continuity in y. Let $\mathcal{V} \subset \mathcal{U} \subset \mathcal{C}^n([a,b])$.

Definition 4.2.4 (Continuous 1st variation). $\delta J[y][\delta y]$ is *continuous* on $\mathcal{U} \times \mathcal{V}$ if it is continuous in each of its arguments. For every $\delta y \in \mathcal{V}$ and $\varepsilon > 0$, there exists $\delta > 0$ such that $|\delta J[y][\delta y] - \delta J[y_0][\delta y]| < \varepsilon$ whenever $\|y - y_0\|_{nsup} < \delta$ where $y, y_0 \in \mathcal{U}$. Furthermore, $\delta J[y][\delta y]$ is a continuous linear functional in δy for every $y \in \mathcal{U}$.

Continuous 2nd variation is defined in the same manner.

Definition 4.2.5 (Continuous 2nd variation). $\delta^2 J[y][\delta y]$ is *continuous* on $\mathcal{U} \times \mathcal{V}$ if it is continuous in each of its arguments. For every $\delta y \in \mathcal{V}$ and $\varepsilon > 0$, there exists $\delta > 0$ such that $|\delta^2 J[y][\delta y] - \delta^2 J[y_0][\delta y]| < \varepsilon$ whenever $\|y - y_0\|_{nsup} < \delta$ where $y, y_0 \in \mathcal{U}$. Furthermore, $\delta^2 J[y][\delta y]$ is a continuous functional in δy for every $y \in \mathcal{U}$.

We denote the class of continuous functionals on $\mathcal{U}$ as $\mathcal{D}(\mathcal{U})$. The set of continuous functionals with continuous 1st variations are notated as $\mathcal{D}^1(\mathcal{U})$. Similarly, $\mathcal{D}^2(\mathcal{U})$ are continuous functionals with continuous 1st and 2nd variations. It is easy to conclude that $\mathcal{D}^2(\mathcal{U}) \subset \mathcal{D}^1(\mathcal{U}) \subset \mathcal{D}(\mathcal{U})$.

We shall assume the continuity of functionals for the remainder of this book. Proving continuity for functionals is usually very complicated. Texts on functional analysis treat this topic in more detail.

4.3 Alternate Variational Formulation

In many texts, calculus of variations is formulated in a manner resembling ordinary calculus. Let $\mathcal{U}$ be the set of possible solutions and $\mathcal{V}$ be the variational space. Let $\delta y(x) = \varepsilon \eta(x)$ where $\eta \in \mathcal{V}$ and ε is a small parameter. We will have $\delta y \in \mathcal{V}$ for sufficiently small ε. The 1st and 2nd order Taylor series expansions of a functional are given by the following lemmas.

Lemma 4.3.1. *If $J \in \mathcal{D}^1(\mathcal{U})$, then for every $y \in \mathcal{U}$ and every $\delta y = \varepsilon \eta \in \mathcal{V}$, we can write*

$$J[y + \varepsilon \eta] = J[y] + \varepsilon\, J_1[y][\eta] + \mu \varepsilon \, \|\eta\|_{nsup} \tag{4.3.1}$$

where $\mu \to 0$ as $\varepsilon \to 0$ and J_1 is a linear functional in η.

Proof. We use the definition of 1st variation in the form of equation (4.2.1).

$$\begin{aligned} J[y + \varepsilon \eta] &= J[y] + \delta J[y][\varepsilon \eta] + \mu_1 \, \|\varepsilon \eta\|_{nsup} \\ &= J[y] + \delta J[y][\varepsilon \eta] + \mu_1 \, |\varepsilon| \, \|\eta\|_{nsup} \end{aligned}$$

$\mu_1 \to 0$ as $\varepsilon \to 0$ because $\|\varepsilon \eta\|_{nsup} \to 0$ when $\varepsilon \to 0$. We can write

$$J[y + \varepsilon \eta] = J[y] + \delta J[y][\varepsilon \eta] + \mu \varepsilon \, \|\eta\|_{nsup}$$

by defining $\mu = \text{sign}(\varepsilon)\, \mu_1$. Note that $\mu \to 0$ as $\varepsilon \to 0$. The factor of ε can be pulled out of the 1st variation because of the linearity property.

$$J[y + \varepsilon \eta] = J[y] + \varepsilon\, \delta J[y][\eta] + \mu \varepsilon \, \|\eta\|_{nsup}$$

Setting $J_1[y][\eta] = \delta J[y][\eta]$ completes the proof. $\square$

Lemma 4.3.2. *If $J \in \mathcal{D}^2(\mathcal{U})$, then for every $y \in \mathcal{U}$ and every $\delta y = \varepsilon \eta \in \mathcal{V}$, we can write*

$$J[y + \varepsilon \eta] = J[y] + \varepsilon\, J_1[y][\eta] + \frac{1}{2}\varepsilon^2\, J_2[y][\eta] + \mu \varepsilon^2 \, \|\eta\|^2_{nsup} \tag{4.3.2}$$

where $\mu \to 0$ as $\varepsilon \to 0$. J_1 is a linear functional in η and J_2 is a quadratic functional in η.

Proof. We use the definition of 2nd variation in the form of equation (4.2.2).

$$\begin{aligned} J[y + \varepsilon \eta] &= J[y] + \delta J[y][\varepsilon \eta] + \delta^2 J[y][\varepsilon \eta] + \mu \, \|\varepsilon \eta\|^2_{nsup} \\ &= J[y] + \varepsilon\, \delta J[y][\eta] + \varepsilon^2\, \delta^2 J[y][\eta] + \mu \varepsilon^2 \, \|\eta\|^2_{nsup} \end{aligned}$$

$\mu \to 0$ as $\varepsilon \to 0$ because $\|\varepsilon \eta\|_{nsup} \to 0$ when $\varepsilon \to 0$. Factors of ε are pulled out of the variations according to the linear and quadratic functional properties. The proof is complete by making the following associations:

$$J_1[y][\eta] = \delta J[y][\eta], \qquad J_2[y][\eta] = 2\, \delta^2 J[y][\eta]$$

$\square$

4.3. ALTERNATE VARIATIONAL FORMULATION

The alternate variational formulation uses derivatives to extract the 1st and 2nd variations of a functional.

Lemma 4.3.3. If $J \in \mathcal{D}^1(\mathcal{U})$, then

$$\frac{d}{d\varepsilon}J[y+\varepsilon\eta]\bigg|_{\varepsilon=0} = J_1[y][\eta] = \delta J[y][\eta], \qquad \forall y \in \mathcal{U}, \forall \eta \in \mathcal{V} \qquad (4.3.3)$$

Proof. Compute the derivative of equation (4.3.1) with respect to ε.

$$\frac{d}{d\varepsilon}J[y+\varepsilon\eta] = J_1[y][\eta] + \mu \, \|\eta\|_{nsup}$$

Recall that $\mu \to 0$ as $\varepsilon \to 0$. Hence, the error term vanishes when setting $\varepsilon = 0$.

$$\frac{d}{d\varepsilon}J[y+\varepsilon\eta]\bigg|_{\varepsilon=0} = J_1[y][\eta] = \delta J[y][\eta]$$

$\square$

Lemma 4.3.4. If $J \in \mathcal{D}^2(\mathcal{U})$, then

$$\frac{d^2}{d\varepsilon^2}J[y+\varepsilon\eta]\bigg|_{\varepsilon=0} = J_2[y][\eta] = 2\,\delta^2 J[y][\eta], \qquad \forall y \in \mathcal{U}, \forall \eta \in \mathcal{V} \qquad (4.3.4)$$

Proof. Compute the 2nd derivative of equation (4.3.2) with respect to ε.

$$\frac{d^2}{d\varepsilon^2}J[y+\varepsilon\eta] = J_2[y][\eta] + 2\mu \, \|\eta\|_{nsup}^2$$

Recall that $\mu \to 0$ as $\varepsilon \to 0$. Hence, the error term vanishes when setting $\varepsilon = 0$.

$$\frac{d^2}{d\varepsilon^2}J[y+\varepsilon\eta]\bigg|_{\varepsilon=0} = J_2[y][\eta] = 2\,\delta^2 J[y][\eta]$$

$\square$

4.4 Integral Functionals

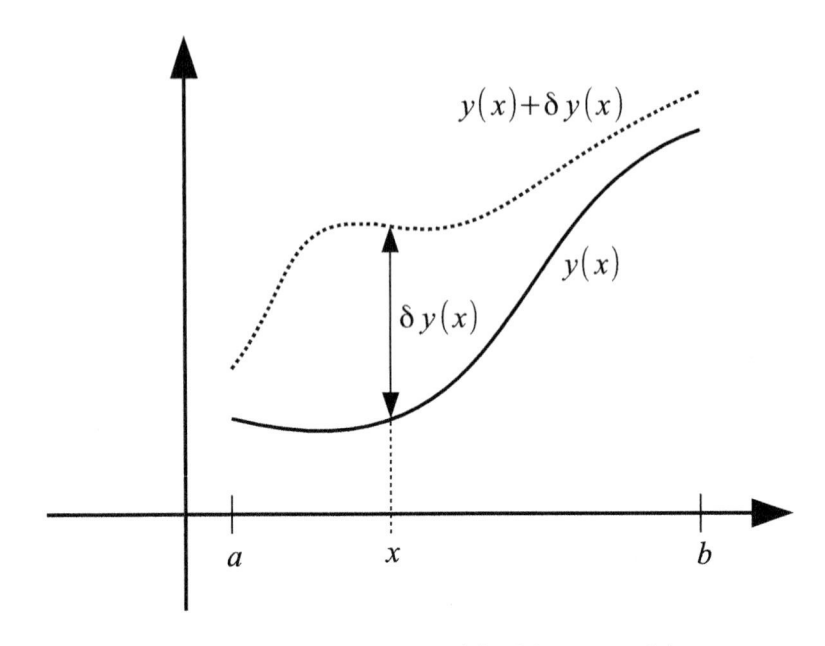

Figure 4.4.1: Variation $\delta y(x)$ of function $y(x)$

We are interested in functionals of the following form:

$$J[y] = \int_a^b F(x, y, y', y'') \, dx \qquad (4.4.1)$$

where F has continuous partial derivatives up to order 2 in each of its arguments.

The variation of F is required to compute the variation of J. For a fixed value of x, we consider the change in F as y changes. Let $\delta y(x)$ be an admissible variation. Then we go from from the value $y(x)$ to $y(x) + \delta y(x)$ inside of F. The corresponding changes in the derivatives are

$$[y(x) + \delta y(x)]' = y'(x) + \delta y'(x)$$
$$[y(x) + \delta y(x)]'' = y''(x) + \delta y''(x)$$

To simplify notation, we write: $\delta y \equiv \delta y(x)$, $\delta y' \equiv \delta y'(x)$, and $\delta y'' \equiv \delta y''(x)$. We emphasize there is no variation in the variable x. The variation occurs in the vertical direction in figure 4.4.1. Hence, we always write $\delta x = 0$.

4.4. INTEGRAL FUNCTIONALS

The goal is to compute the value of $F(x, y + \delta y, y' + \delta y', y'' + \delta y'')$. The continuity of F with respect to its partial derivatives means that it has a Taylor series expansion. We introduce $\mathbf{u} = [x, y, y', y'']^T$ and write $F(\mathbf{u}) \equiv F(x, y, y', y'')$. The 1st order expansion of F is provided by equation (3.2.12).

$$F(\mathbf{u} + \delta \mathbf{u}) = F(\mathbf{u}) + \nabla F(\mathbf{u}) \cdot \delta \mathbf{u} + r(\delta \mathbf{u})$$

where $\delta \mathbf{u} = [\delta x, \delta y, \delta y', \delta y'']^T$ and $r(\delta \mathbf{u})$ is $o(\|\delta \mathbf{u}\|_2)$ as $\delta \mathbf{u} \to \mathbf{0}$. Using variational notation and noting that $\delta x = 0$, we have

$$\delta F \equiv \nabla F(\mathbf{u}) \cdot \delta \mathbf{u} = \frac{\partial}{\partial y} F(\mathbf{u})\, \delta y + \frac{\partial}{\partial y'} F(\mathbf{u})\, \delta y' + \frac{\partial}{\partial y''} F(\mathbf{u})\, \delta y''$$

$$F(\mathbf{u} + \delta \mathbf{u}) = F(\mathbf{u}) + \delta F + r(\delta \mathbf{u})$$

Using this result, the increment in J is computed.

$$\begin{aligned}
J[y + \delta y] &= \int_a^b F(\mathbf{u} + \delta \mathbf{u})\, dx \\
&= \int_a^b [F(\mathbf{u}) + \delta F + r(\delta \mathbf{u})]\, dx \\
&= \int_a^b F(\mathbf{u})\, dx + \int_a^b \delta F\, dx + \int_a^b r(\delta \mathbf{u})\, dx \\
&= J[y] + \int_a^b \delta F\, dx + \int_a^b r(\delta \mathbf{u})\, dx
\end{aligned}$$

We claim that the 1st variation of J is given by

$$\delta J \equiv \delta J[y][\delta y] = \int_a^b \delta F\, dx \qquad (4.4.2)$$

δJ is linear in δy because the integral of the linear functional δF is also linear.

The final step in verifying that equation (4.4.2) is the 1st variation is to show the error term has the correct form. One problem is that every component of $\delta \mathbf{u}$ is a function of x. We cannot simply pull $r(\mathbf{u})$ out of the integral. Consider the following inequalities:

$$\left| \int_a^b r(\delta \mathbf{u})\, dx \right| \leq \int_a^b |r(\delta \mathbf{u})|\, dx < \int_a^b \varepsilon(x)\, \|\delta \mathbf{u}\|_2\, dx$$

$\varepsilon(x) \to 0$ as $\delta \mathbf{u} \to 0$ for every $x \in I = [a, b]$ because $r(\delta \mathbf{u})$ is $o(\|\delta \mathbf{u}\|_2)$. Note that $\varepsilon(x) > 0$ according to the definition of little o.

We need to write the error in terms $\|\delta y\|_{2sup}$. The next inequality arises by maximizing the square of each component of $\delta\mathbf{u}$ over $[a,b]$.

$$\int_a^b \varepsilon(x) \|\delta\mathbf{u}\|_2 \, dx = \int_a^b \varepsilon(x) \sqrt{(\delta y)^2 + (\delta y')^2 + (\delta y'')^2} \, dx$$

$$\leq \int_a^b \varepsilon(x) \sqrt{\max_{x \in I}(\delta y)^2 + \max_{x \in I}(\delta y')^2 + \max_{x \in I}(\delta y'')^2} \, dx$$

Another inequality follows by the triangle inequality.

$$\int_a^b \varepsilon(x) \|\delta\mathbf{u}\|_2 \, dx \leq \int_a^b \varepsilon(x) \left[\sqrt{\max_{x \in I}(\delta y)^2} + \sqrt{\max_{x \in I}(\delta y')^2} + \sqrt{\max_{x \in I}(\delta y'')^2} \right] dx$$

$$\leq \int_a^b \varepsilon(x) \left[\max_{x \in I} |\delta y| + \max_{x \in I} |\delta y'| + \max_{x \in I} |\delta y''| \right] dx$$

$$\leq \int_a^b \varepsilon(x) \|\delta y\|_{2sup} \, dx$$

$$\leq \|\delta y\|_{2sup} \int_a^b \varepsilon(x) \, dx$$

All four right hand sides above are equal. We now maximize $\varepsilon(x)$ to get

$$\|\delta y\|_{2sup} \int_a^b \varepsilon(x) \, dx \leq \|\delta y\|_{2sup} \int_a^b \|\varepsilon\|_{\sup} \, dx = (b-a) \|\varepsilon\|_{\sup} \|\delta y\|_{2sup}$$

Let $\tilde{\varepsilon} = (b-a) \|\varepsilon\|_{\sup}$ and we have our final inequality

$$\left| \int_a^b r(\delta\mathbf{u}) \, dx \right| < \tilde{\varepsilon} \|\delta y\|_{2sup}$$

$\tilde{\varepsilon} \to 0$ as $\|\delta y\|_{2nsup} \to 0$ because $\varepsilon(x) \to 0$ as $\delta\mathbf{u}(x) \to \mathbf{0}$ for every $x \in I$. The error term is the correct form and order for equation (4.4.2) to be the 1st variation of J.

Suppose functional J contained the 1st derivative but not the 2nd derivative. Then the previous results remain intact except that $\mathbf{u} = [x, y, y', 0]^T$, $\delta\mathbf{u} = [\delta x, \delta y, \delta y', 0]^T$, and we replace $\|\delta y\|_{2sup}$ with $\|\delta y\|_{1sup}$.

Another observation concerning the 1st variation is that the variation operator commutes with integration. This fact is easily seen by writing

$$\delta J = \delta(J) = \delta \int_a^b F \, dx = \int_a^b \delta F \, dx \qquad (4.4.3)$$

This property is used to compute variations of functionals with different integrands. We thus avoid reproducing the derivations in this section.

4.4. INTEGRAL FUNCTIONALS

Equation (3.4.1) provides the 2nd order expansion of F.

$$F(\mathbf{u} + \delta\mathbf{u}) = F(\mathbf{u}) + \nabla F(\mathbf{u}) \cdot \delta\mathbf{u} + \frac{1}{2}(\delta\mathbf{u})^T \mathbf{H}(\mathbf{u})\,\delta\mathbf{u} + r(\delta\mathbf{u})$$

where $r(\delta\mathbf{u})$ is $o(\|\delta\mathbf{u}\|_2^2)$ as $\delta\mathbf{u} \to 0$. The 2nd variation of F is

$$\delta^2 F \equiv \frac{1}{2}(\delta\mathbf{u})^T \mathbf{H}(\mathbf{u})\,\delta\mathbf{u}$$

with the Hessian matrix

$$\mathbf{H}(\mathbf{u}) = \begin{bmatrix} \frac{\partial^2}{\partial x^2}F(\mathbf{u}) & \frac{\partial^2}{\partial x \partial y}F(\mathbf{u}) & \frac{\partial^2}{\partial x \partial y'}F(\mathbf{u}) & \frac{\partial^2}{\partial x \partial y''}F(\mathbf{u}) \\ \frac{\partial^2}{\partial x \partial y}F(\mathbf{u}) & \frac{\partial^2}{\partial y^2}F(\mathbf{u}) & \frac{\partial^2}{\partial y \partial y'}F(\mathbf{u}) & \frac{\partial^2}{\partial y \partial y''}F(\mathbf{u}) \\ \frac{\partial^2}{\partial x \partial y'}F(\mathbf{u}) & \frac{\partial^2}{\partial y \partial y'}F(\mathbf{u}) & \frac{\partial^2}{\partial (y')^2}F(\mathbf{u}) & \frac{\partial^2}{\partial y' \partial y''}F(\mathbf{u}) \\ \frac{\partial^2}{\partial x \partial y''}F(\mathbf{u}) & \frac{\partial^2}{\partial y \partial y''}F(\mathbf{u}) & \frac{\partial^2}{\partial y' \partial y''}F(\mathbf{u}) & \frac{\partial^2}{\partial (y'')^2}F(\mathbf{u}) \end{bmatrix}$$

Setting $\delta x = 0$ and performing the matrix multiplication gives

$$\begin{aligned} \delta^2 F &= \frac{1}{2}\left[\frac{\partial^2}{\partial y^2}F(\mathbf{u})(\delta y)^2 + \frac{\partial^2}{\partial (y')^2}F(\mathbf{u})(\delta y')^2 + \frac{\partial^2}{\partial (y'')^2}F(\mathbf{u})(\delta y'')^2\right] \\ &+ \frac{\partial^2}{\partial y \partial y'}F(\mathbf{u})\,\delta y\,\delta y' + \frac{\partial^2}{\partial y' \partial y''}F(\mathbf{u})\,\delta y'\,\delta y'' + \frac{\partial^2}{\partial y \partial y''}F(\mathbf{u})\,\delta y\,\delta y'' \end{aligned}$$

$$F(\mathbf{u} + \delta\mathbf{u}) = F(\mathbf{u}) + \delta F + \delta^2 F + r(\delta\mathbf{u})$$

The increment in J is computed as

$$\begin{aligned} J[y + \delta y] &= \int_a^b F(\mathbf{u} + \delta\mathbf{u})\,dx \\ &= \int_a^b \left[F(\mathbf{u}) + \delta F + \delta^2 F + r(\delta\mathbf{u})\right]dx \\ &= \int_a^b F(\mathbf{u})\,dx + \int_a^b \delta F\,dx + \int_a^b \delta^2 F\,dx + \int_a^b r(\delta\mathbf{u})\,dx \\ &= J[y] + \delta J + \int_a^b \delta^2 F\,dx + \int_a^b r(\delta\mathbf{u})\,dx \end{aligned}$$

The 2nd variation of J is

$$\delta^2 J \equiv \delta^2 J[y][\delta y] = \int_a^b \delta^2 F\,dx \qquad (4.4.4)$$

$\delta^2 J$ is a quadratic functional because δF is a quadratic functional of $\delta\mathbf{u}$.

The error term for the 2nd variation must be analyzed. We start with the inequality

$$\left|\int_a^b r(\delta\mathbf{u})\,dx\right| \leq \int_a^b |r(\delta\mathbf{u})|\,dx < \int_a^b \varepsilon(x)\,\|\delta\mathbf{u}\|_2^2\,dx$$

$\varepsilon(x) \to 0$ as $\delta\mathbf{u} \to \mathbf{0}$ for every $x \in I = [a,b]$ because $r(\delta\mathbf{u})$ is $o(\|\delta\mathbf{u}\|_2^2)$.

Transformation to $\|\delta y\|_{2sup}$ begins with maximizing the squares of the components of $\delta\mathbf{u}$.

$$\int_a^b \varepsilon(x)\,\|\delta\mathbf{u}\|_2^2\,dx \leq \int_a^b \varepsilon(x)\left[\max_{x\in I}(\delta y)^2 + \max_{x\in I}(\delta y')^2 + \max_{x\in I}(\delta y'')^2\right]dx$$

The right hand side is obviously less than or equal to the next right hand side below.

$$\begin{aligned}
\int_a^b \varepsilon(x)\,\|\delta\mathbf{u}\|_2^2\,dx &\leq \int_a^b \varepsilon(x)\left[\max_{x\in I}|\delta y| + \max_{x\in I}|\delta y'| + \max_{x\in I}|\delta y''|\right]^2 dx \\
&\leq \int_a^b \varepsilon(x)\,\|\delta y\|_{2sup}^2\,dx \\
&\leq \|\delta y\|_{2sup}^2 \int_a^b \varepsilon(x)\,dx
\end{aligned}$$

The last three right hand sides are equal. Maximizing $\varepsilon(x)$ gives

$$\|\delta y\|_{2sup}^2 \int_a^b \varepsilon(x)\,dx \leq \|\delta y\|_{2sup}^2 \int_a^b \|\varepsilon\|_{\sup}\,dx = (b-a)\,\|\varepsilon\|_{\sup}\,\|\delta y\|_{2sup}^2$$

Set $\tilde{\varepsilon} = (b-a)\,\|\varepsilon\|_{\sup}$ to obtain

$$\left|\int_a^b r(\delta\mathbf{u})\,dx\right| < \tilde{\varepsilon}\,\|\delta y\|_{2sup}^2$$

$\tilde{\varepsilon} \to 0$ as $\|\delta y\|_{2nsup} \to 0$ because $\varepsilon(x) \to 0$ as $\delta\mathbf{u}(x) \to \mathbf{0}$ for every $x \in I$. This inequality provides the correct order for the error term in equation (4.4.4) to be the 2nd variation of J.

If the functional J contained the 1st derivative but not the 2nd derivative, then use $\mathbf{u} = [x, y, y', 0]^T$, $\delta\mathbf{u} = [\delta x, \delta y, \delta y', 0]^T$, and $\|\delta y\|_{1sup}$.

The 2nd variation operator also commutes with integration.

$$\delta^2 J = \delta^2(J) = \delta^2 \int_a^b F\,dx = \int_a^b \delta^2 F\,dx \qquad (4.4.5)$$

4.5 Local Minimums

Consider the closed interval $I = [a, b]$. The concept of an open ball in the set $\mathcal{U} \subset \mathcal{C}^n(I)$ is provided by the following definition.

Definition 4.5.1 (Open Ball). The *open ball* $B(y_0, \delta)$ centered at y_0 with radius δ is the set of all functions $y \in \mathcal{U}$ such that $\|y - y_0\|_{nsup} < \delta$.

The functions y^* which cause the 1st variation δJ to vanish are called stationary points. These points play an important role in the calculus of variations. Stationary points are often sufficient for certain applications. It may be difficult, or even impossible, to compute the minimums of some functionals.

Definition 4.5.2 (Stationary Point). y^* is the location of a *stationary point* of the functional $J[y]$ if $\delta J[y^*][\delta y] = 0$ for all admissible variations δy.

Being a stationary point is a necessary condition for a local minimum.

Definition 4.5.3 (Local Minimum). y^* is the location of a *local minimum* of $J[y]$ if there exists $\delta > 0$ such that $J[y^*] \leq J[y]$ for all $y \in B(y^*, \delta)$.

We again only concern ourselves with local minimums. The problem of global minimization of functionals is difficult and is not discussed in this book.

Theorem 4.5.1 (1st Order Necessary Condition). *If y^* is the location of a local minimum of $J[y]$ and $J \in \mathcal{D}^1(\mathcal{U})$, then $\delta J[y^*][\delta y] = 0$ for all admissible variations δy.*

Proof. The proof is by contradiction. Suppose that $\delta J[y^*][\delta y] \neq 0$ for some admissible variation $\delta y \neq 0$. The definition of 1st variation states that

$$J[y^* + \alpha\,\delta y] = J[y^*] + \delta J[y^*][\alpha\,\delta y] + \varepsilon\,\|\alpha\,\delta y\|_{nsup}$$

for all $\alpha \in [0, 1]$. The error term approaches 0 faster than the 1st variation. If this was not true, then lemma 4.2.2 implies the 1st variation would vanish. Thus, there exists $\delta > 0$ such that $|\varepsilon|\,\|\alpha\,\delta y\|_{nsup} < \frac{1}{2}|\delta J[y^*][\alpha\,\delta y]|$ whenever $\|\alpha\,\delta y\|_{nsup} < \delta$. Set $\bar{\alpha} = \delta/\|\delta y\|_{nsup}$ as the upper bound for the size of α. If the 1st variation is negative, then

$$J[y^* + \alpha\,\delta y] \leq J[y^*] + \frac{1}{2}\delta J[y^*][\alpha\,\delta y] = J[y^*] + \frac{1}{2}\alpha\,\delta J[y^*][\delta y] < J[y^*]$$

for all $\alpha \in (0, \bar{\alpha})$. If the 1st variation is positive, then choose $\hat{\alpha} \in (0, \bar{\alpha}]$ small enough so that $y^* - \alpha\,\delta y \in \mathcal{U}$ for all $\alpha \in (0, \hat{\alpha})$. We now have

$$J[y^* - \alpha\,\delta y] \leq J[y^*] + \frac{1}{2}\delta J[y^*][-\alpha\,\delta y] = J[y^*] - \frac{1}{2}\alpha\,\delta J[y^*][\delta y] < J[y^*]$$

or all $\alpha \in (0, \hat{\alpha})$. In either case we contradict that y^* is a local minimum. Thus the 1st variation must be 0. □

Theorem 4.5.2 (2nd Order Necessary Condition). *If y^* is the location of a local minimum of $J[y]$ and $J \in \mathcal{D}^2(\mathcal{U})$, then $\delta J[y^*][\delta y] = 0$ and $\delta^2 J[y^*][\delta y] \geq 0$ for all admissible variations δy.*

Proof. The proof is by contradiction. Suppose that $\delta^2 J[y^*][\delta y] < 0$ for some admissible variation $\delta y \neq 0$. The 1st variation vanishes due to the 1st order necessary condition. The definition of 2nd variation at y^* gives

$$J[y^* + \alpha\,\delta y] = J[y^*] + \delta^2 J[y^*][\alpha\,\delta y] + \varepsilon \left\| \alpha\,\delta y \right\|_{nsup}^2$$

for all $\alpha \in [0,1]$. The error term approaches 0 faster than the 2nd variation. Otherwise, lemma 4.2.5 states that the 2nd variation vanishes. Thus, there exists $\delta > 0$ such that $|\varepsilon|\,\|\alpha\,\delta y\|_{nsup}^2 < \frac{1}{2} |\delta^2 J[y^*][\alpha\,\delta y]|$ for all $\|\alpha\,\delta y\|_{nsup} < \delta$. Set $\bar{\alpha} = \delta/\|\delta y\|_{nsup}$ as the upper bound for the size of α.

$$J[y^* + \alpha\,\delta y] \leq J[y^*] + \frac{1}{2}\delta^2 J[y^*][\alpha\,\delta y] = J[y^*] + \frac{1}{2}\alpha^2\,\delta^2 J[y^*][\delta y] < J[y^*]$$

for all $\alpha \in (0, \bar{\alpha})$. This contradicts that y^* is a local minimum. Thus the 2nd variation must be non-negative. $\square$

Theorem 4.5.3 (2nd Order Sufficient Condition). *If $J \in \mathcal{D}^2(\mathcal{U})$, $\delta J[y^*][\delta y] = 0$, and $\delta^2 J[y^*][\delta y] > 0$ for all admissible variations $\delta y \neq 0$, then y^* is the location of a local minimum of $J[y]$.*

Proof. By the same reasoning in the proof of the 2nd order necessary condition, there exists $\delta > 0$ such that $|\varepsilon|\,\|\delta y\|_{nsup}^2 < \frac{1}{2}\delta^2 J[y^*][\delta y]$ for all $\|\delta y\|_{nsup} < \delta$.

$$J[y^* + \delta y] = J[y^*] + \delta^2 J[y^*][\delta y] + \varepsilon \|\delta y\|_{nsup}^2 > J[y^*] + \frac{1}{2}\delta^2 J[y^*][\delta y] > J[y^*]$$

The strict inequality is due to the fact $\delta y \neq 0$. Thus y^* is the location of a local minimum. $\square$

Proving the positivity of the 2nd variation for all admissible variations is often difficult. Gelfand and Fomin [9] extensively discuss this issue.

4.6 Fundamental Lemmas

This section provides essential lemmas for the calculus of variations. They are used to prove several results which appear in the remainder of this book. Let $I = [a, b]$ and $\mathcal{C}_0^n(I) = \{v \in \mathcal{C}^n(I) \,|\, v(a) = v(b) = 0\}$. For any chosen order of differentiability $n \geq 0$, we have the Fundamental Lemma of Calculus of Variations.

Lemma 4.6.1 (Fundamental Lemma of Calculus of Variations). *Suppose $f \in \mathcal{C}(I)$ and*

$$\int_a^b f(x)\, \eta(x)\, dx = 0, \qquad \forall \eta \in \mathcal{C}_0^n(I) \tag{4.6.1}$$

then $f(x) = 0$ for all $x \in I$.

Proof. The proof is by contradiction. Suppose that $f(x) \neq 0$. Then the continuity of $f(x)$ implies there exists an open interval $K = (x_0, x_1) \subset I$ where $f(x)$ is nonzero. We will assume that $f(x) > 0$ on K. The proof for $f(x) < 0$ on K is similar and will be omitted.

Now define a particular $\eta(x)$ using

$$\eta(x) = \begin{cases} (x - x_0)^m (x - x_1)^m & \text{if } x \in K \\ 0 & \text{if } x \notin K \end{cases}$$

where $m = 2(n+1)$. It is obvious that $\eta(x)$ is positive on K, continuous on I, and vanishes at a and b. We must show that $\eta \in \mathcal{C}_0^n(I)$. Let $u(x) = (x - x_0)^m$ and $v(x) = (x - x_1)^m$. Then the derivatives $u^{(k)}(x) = m(m-1)\cdots(m-k+1)(x - x_0)^{m-k}$ and $v^{(k)}(x) = m(m-1)\cdots(m-k+1)(x - x_1)^{m-k}$ are continuous on K. The k-th derivative of $\eta(x)$ is computed using successive product rules.

$$\eta^{(k)}(x) = \begin{cases} \sum_{j=0}^{k} \binom{k}{j} u^{(k-j)}(x)\, v^{(j)}(x) & \text{if } x \in K \\ 0 & \text{if } x \notin K \end{cases}$$

where $k = 1, \ldots, n$ and $k < m$. Notice that $u^{(k-j)}(x_0) = 0$ and $v^{(j)}(x_1) = 0$ for $j = 0, \ldots, k$. This ensures that $\eta^{(k)}(x_0) = 0$ and $\eta^{(k)}(x_1) = 0$. Thus $\eta^{(k)}(x)$ is continuous on I and $\eta \in \mathcal{C}_0^n(I)$.

Returning to the integral, we have

$$\int_a^b f(x)\, \eta(x)\, dx = \int_{x_0}^{x_1} f(x)\, \eta(x)\, dx > 0$$

This contradicts the hypothesis (4.6.1) of the theorem. Thus $f(x) = 0$ on I. $\square$

Lemma 4.6.2. *Suppose $f \in C(I)$ and*

$$\int_a^b f(x)\eta'(x)\,dx = 0, \qquad \forall \eta \in C_0^1(I) \tag{4.6.2}$$

then $f(x) = c$ for all $x \in I$, where c is a constant.

Proof. Let c be the average value of f on I.

$$c = \frac{1}{b-a}\int_a^b f(x)\,dx$$

Then we have

$$\int_a^b [f(x) - c]\,dx = 0 \tag{4.6.3}$$

Now define a particular function $\eta \in C_0^1(I)$ using

$$\eta(x) = \int_a^x [f(s) - c]\,ds \tag{4.6.4}$$

Notice that

$$\eta'(x) = f(x) - c \tag{4.6.5}$$

Now compute

$$\begin{aligned}
\int_a^b [f(x) - c]\eta'(x)\,dx &= \int_a^b f(x)\eta'(x)\,dx - \int_a^b c\,\eta'(x)\,dx \\
&= \int_a^b f(x)\eta'(x)\,dx - c\int_a^b \eta'(x)\,dx \\
&= \int_a^b f(x)\eta'(x)\,dx - c\int_a^b [f(x) - c]\,dx
\end{aligned}$$

The first term is zero because of the hypothesis (4.6.2). The second term is also zero because of equation (4.6.3).

$$\int_a^b [f(x) - c]\eta'(x)\,dx = 0 \tag{4.6.6}$$

Substituting the expression (4.6.5) for $\eta'(x)$ into equation (4.6.6), we now have

$$\int_a^b [f(x) - c]^2\,dx = 0$$

Since $f(x)$ is continuous on I, this is only possible if $f(x) = c$ on I. □

4.6. FUNDAMENTAL LEMMAS

Lemma 4.6.3. *Suppose $f, g \in \mathcal{C}(I)$ and*

$$\int_a^b [f(x)\,\eta(x) + g(x)\,\eta'(x)]\,dx = 0, \qquad \forall \eta \in \mathcal{C}_0^1(I) \tag{4.6.7}$$

then $g \in \mathcal{C}^1(I)$ and $g'(x) = f(x)$ for all $x \in I$.

Proof. First define $F \in \mathcal{C}^1(I)$ in the following way:

$$F(x) = \int_a^x f(s)\,ds \tag{4.6.8}$$

Integrating the first term in equation (4.6.7) by parts.

$$\int_a^b f(x)\,\eta(x)\,dx \;=\; [F(x)\,\eta(x)]\Big|_a^b - \int_a^b F(x)\,\eta'(x)\,dx$$

The boundary terms vanish because $\eta \in \mathcal{C}_0^1(I)$.

$$\int_a^b f(x)\,\eta(x)\,dx = -\int_a^b F(x)\,\eta'(x)\,dx \tag{4.6.9}$$

We substitute the right hand side of equation (4.6.9) into equation (4.6.7) and factor out $\eta'(x)$.

$$\int_a^b [-F(x) + g(x)]\eta'(x)\,dx = 0, \qquad \forall \eta \in \mathcal{C}_0^1(I) \tag{4.6.10}$$

We apply lemma 4.6.2 to equation (4.6.10) to conclude that

$$-F(x) + g(x) = c, \qquad \forall x \in I$$

where c is a constant. Taking the derivative of this expression, we have

$$\begin{aligned}
-F'(x) + g'(x) &= 0 \\
g'(x) &= F'(x) \\
g'(x) &= f(x)
\end{aligned}$$

Furthermore, $g \in \mathcal{C}^1(I)$ because $f \in \mathcal{C}(I)$. $\qquad \square$

Lemma 4.6.4. *Suppose $f \in \mathcal{C}(I)$ and*

$$\int_a^b f(x)\, \eta''(x)\, dx = 0, \qquad \forall \eta \in \mathcal{V} \qquad (4.6.11)$$

where $\mathcal{V} = \{\eta \in \mathcal{C}^2(I) \mid \eta(a) = \eta(b) = \eta'(a) = \eta'(b) = 0\}$. Then $f(x) = c_0 + c_1 x$ for all $x \in I$ where c_0 and c_1 are constants.

Proof. First, compute the constants c_0 and c_1 by solving the following system of equations:

$$\int_a^b [f(x) - (c_0 + c_1 x)]\, dx = 0$$

$$\int_a^b \int_a^x [f(s) - (c_0 + c_1 s)]\, ds\, dx = 0$$

We define a particular function $\eta \in \mathcal{V}$ as

$$\eta(x) = \int_a^x \int_a^s [f(\xi) - (c_0 + c_1 \xi)]\, d\xi\, ds \qquad (4.6.12)$$

The first two derivatives of this function are

$$\eta'(x) = \int_a^x [f(s) - (c_0 + c_1 s)]\, ds \qquad (4.6.13)$$

$$\eta''(x) = f(s) - (c_0 + c_1 x) \qquad (4.6.14)$$

The computation of c_0 and c_1 ensures that $\eta(a) = \eta(b) = \eta'(a) = \eta'(b) = 0$ and hence $\eta \in \mathcal{V}$.

4.6. FUNDAMENTAL LEMMAS

We compute the following integral:

$$\begin{aligned}
\int_a^b [f(x) - (c_0 + c_1 x)]\, \eta''(x)\, dx &= \int_a^b f(x) \eta''(x)\, dx - c_0 \int_a^b \eta''(x)\, dx \\
&\quad - c_1 \int_a^b x\, \eta''(x)\, dx \\
&= \int_a^b f(x) \eta''(x)\, dx - [c_0 \eta'(x)]\big|_a^b \\
&\quad - [c_1 x\, \eta'(x)]\big|_a^b + c_1 \int_a^b \eta'(x)\, dx \\
&= \int_a^b f(x) \eta''(x)\, dx - 0 - 0 + [c_1 \eta(x)]\big|_a^b \\
&= \int_a^b f(x) \eta''(x)\, dx + 0 \\
&= \int_a^b f(x) \eta''(x)\, dx
\end{aligned}$$

The integral, on the right hand side of the last line, vanishes because of equation (4.6.11).

$$\int_a^b [f(x) - (c_0 + c_1 x)]\, \eta''(x)\, dx = 0 \qquad (4.6.15)$$

Now substitute $\eta''(x)$ from equation (4.6.14) into equation (4.6.15).

$$\int_a^b [f(x) - (c_0 + c_1 x)]^2\, dx = 0$$

The expression in brackets is a continuous function on I. The only way this integral vanishes is if

$$f(x) = c_0 + c_1 x, \qquad \forall x \in I$$

□

Lemma 4.6.5. Suppose $f, g, h \in \mathcal{C}^1(I)$ and

$$\int_a^b [f(x)\,\eta(x) + g(x)\,\eta'(x) + h(x)\,\eta''(x)]\,dx = 0, \qquad \forall \eta \in \mathcal{V} \qquad (4.6.16)$$

where $\mathcal{V} = \{\eta \in \mathcal{C}^2(I) \mid \eta(a) = \eta(b) = \eta'(a) = \eta'(b) = 0\}$. Then we have

$$f(x) - g'(x) + h''(x) = 0, \qquad \forall x \in I \qquad (4.6.17)$$

and $h \in \mathcal{C}^2(I)$.

Proof. We define the following anti-derivatives:

$$F(x) = \int_a^x f(s)\,ds$$

$$\tilde{F}(x) = \int_a^x F(s)\,ds$$

$$G(x) = \int_a^x g(s)\,ds$$

Note that $F, \tilde{F}, G \in \mathcal{C}^2(I)$.

The first term in (4.6.16) is integrated by parts twice. We also use the fact that $\eta(x)$ and $\eta'(x)$ vanish at the $x = a$ and $x = b$.

$$\int_a^b f(x)\,\eta(x)\,dx = [F(x)\,\eta(x)]\Big|_a^b - \int_a^b F(x)\,\eta'(x)\,dx$$

$$= -\int_a^b F(x)\,\eta'(x)\,dx$$

$$-\int_a^b F(x)\,\eta'(x)\,dx = -\left[\tilde{F}(x)\,\eta'(x)\right]\Big|_a^b + \int_a^b \tilde{F}(x)\,\eta''(x)\,dx$$

$$= \int_a^b \tilde{F}(x)\,\eta''(x)\,dx$$

4.6. FUNDAMENTAL LEMMAS

The second term in (4.6.16) is also integrated by parts.

$$\begin{aligned}\int_a^b g(x)\,\eta'(x)\,dx &= [G(x)\,\eta'(x)]\big|_a^b - \int_a^b G(x)\,\eta''(x)\,dx \\ &= -\int_a^b G(x)\,\eta''(x)\,dx\end{aligned}$$

These results are substituted into equation (4.6.16).

$$\int_a^b \left[\tilde{F}(x) - G(x) + h(x)\right]\eta''(x)\,dx = 0, \qquad \forall \eta \in \mathcal{V} \tag{4.6.18}$$

Lemma 4.6.4 is applied to equation (4.6.18).

$$\tilde{F}(x) - G(x) + h(x) = c_0 + c_1 x, \qquad \forall x \in I$$

where c_0 and c_1 are constants. This equation is differentiated twice to produce equation (4.6.17).

$$f(x) - g'(x) + h''(x) = 0, \qquad \forall x \in I$$

The fact that $f, g \in \mathcal{C}^1(I)$ implies that $h \in \mathcal{C}^2(I)$. □

Chapter 5

Function with 1st Derivative

5.1 Fixed End Points

The previous chapter provided necessary conditions for determining if $y(x)$ is a minimum of the functional

$$J[y] = \int_a^b F(x, y, y') \, dx \qquad (5.1.1)$$

The resulting conditions were in integral form and are difficult to use. We wish to transform the problem to differential form. The transformation is accomplished using integration by parts. The procedure requires that F have a sufficient order of continuity of its partial derivatives. For the 1st variation, we assume that F has continuous derivatives up to order 2 in each of its arguments. Continuous derivatives up to order 3 will be required when investigating the 2nd variation.

The function $y(x)$ has fixed values at the endpoints. Let $I = [a, b]$ be the domain of $y(x)$ and define $\mathcal{U} = \{ y \in \mathcal{C}^1(I) \,|\, y(a) = y_a \text{ and } y(b) = y_b \}$ as the set of possible solutions. We immediately notice that $\mathcal{U}$ is not a subspace of $\mathcal{C}^1(I)$. The lack of closure for vector addition is easily demonstrated. If $y_1, y_2 \in \mathcal{U}$, then $y_1(a) + y_2(a) = 2y_a \neq y_a$ does not satisfy the boundary condition at a. The same problem occurs at the endpoint b. Hence $y_1 + y_2 \notin \mathcal{U}$ and thus $\mathcal{U}$ is not a vector space.

The local minimization problem is formally stated as

$$y^* = \arg \operatorname*{local\,min}_{y \in \mathcal{U}} J[y] \qquad (5.1.2)$$

Calculus of variations is used to solve problem (5.1.2). The techniques and results of the previous chapter will be applied to this problem.

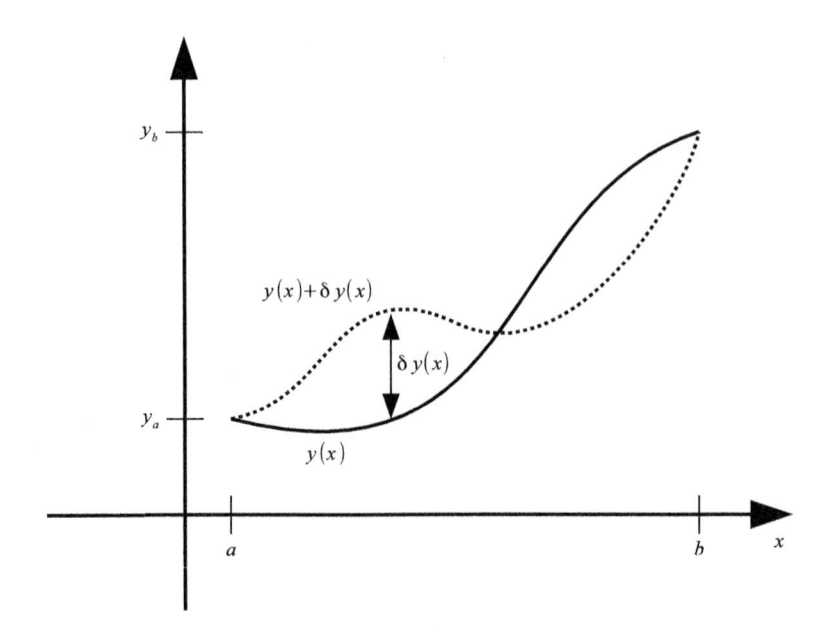

Figure 5.1.1: Variation $\delta y(x)$ of function $y(x)$

The 1st order necessary condition will lead to the Euler differential equation. A 2nd order necessary condition is provided by the Legendre condition. Sufficient conditions are more complicated and are not discussed in this book. Extensive information on sufficient conditions is found in [9].

The set $\mathcal{V}$ of admissible variations must be defined. Any variation δy must have at least the same order of continuous differentiability as the function y. This requirement implies $\mathcal{V} \subset \mathcal{C}^1(I)$. In order for $y + \delta y \in \mathcal{U}$, there can be no variation at the endpoints. This is seen in figure 5.1.1. We define the set of admissible variations as: $\mathcal{V} = \mathcal{C}_0^1(I)$.

The 1st variation of the functional J, defined by equation (5.1.1), can be computed directly by the results of the previous chapter. We demonstrate an alternative derivation based on using the commutative property (4.4.3) of variation. Let $F \equiv F(x, y, y')$.

$$\delta J = \delta \int_a^b F \, dx = \int_a^b \delta F \, dx$$

Computing the variation of δF gives

$$\delta J = \int_a^b \left[\frac{\partial F}{\partial y} \delta y + \frac{\partial F}{\partial y'} \delta y' \right] dx \qquad (5.1.3)$$

This simplified procedure is demonstrated in many books on calculus of variations.

5.1. FIXED END POINTS

Theorem 5.1.1 (Euler Differential Equation). *Consider the closed interval $I = [a, b]$, the set of functions $\mathcal{U} = \{y \in \mathcal{C}^1(I) \,|\, y(a) = y_a \text{ and } y(b) = y_b\}$, and the functional $J : \mathcal{U} \to \mathbb{R}$ defined by*

$$J[y] = \int_a^b F(x, y, y') \, dx \tag{5.1.4}$$

where F has continuous partial derivatives up to order 2. A necessary condition for a local minimum of J over $\mathcal{U}$ is that the Euler differential equation be satisfied.

$$\frac{\partial F}{\partial y} - \frac{d}{dx}\frac{\partial F}{\partial y'} = 0 \tag{5.1.5}$$

Proof. Lemma 4.6.3, applied to equation (5.1.3) with $\delta J = 0$, allows us to immediately conclude

$$\frac{d}{dx}\frac{\partial F}{\partial y'} = \frac{\partial F}{\partial y}$$

However, we provide an alternate proof which is found in most texts.

Integration by parts can be performed on the second term in brackets in equation (5.1.3).

$$\begin{aligned}
\delta J &= \int_a^b \left[\frac{\partial F}{\partial y} \delta y + \frac{\partial F}{\partial y'} \delta y' \right] dx \\
&= \int_a^b \frac{\partial F}{\partial y} \delta y \, dx + \int_a^b \frac{\partial F}{\partial y'} \delta y' \, dx \\
&= \int_a^b \frac{\partial F}{\partial y} \delta y \, dx - \int_a^b \left[\frac{d}{dx}\frac{\partial F}{\partial y'} \right] \delta y \, dx + \left[\frac{\partial F}{\partial y'} \delta y \right]_a^b
\end{aligned}$$

The boundary term is zero because $\delta y(a) = \delta y(b) = 0$. We recombine the integrals.

$$\delta J = \int_a^b \left[\frac{\partial F}{\partial y} - \frac{d}{dx}\frac{\partial F}{\partial y'} \right] \delta y \, dx$$

The term in brackets is continuous because of lemma 4.6.3. We apply the Fundamental Lemma 4.6.1 to the requirement that $\delta J = 0$ for all admissible variations δy. The term in brackets must vanish.

$$\frac{\partial F}{\partial y} - \frac{d}{dx}\frac{\partial F}{\partial y'} = 0$$

□

The solution to the Euler differential equation (5.1.5) provides a function $y(x)$ which may be a local minimum of the functional J. At the very least, the solution $y(x)$ is a stationary value of J and this may be sufficient for many applications.

The solution space for the minimization problem (5.1.2) only requires continuous 1st derivatives. It can be shown that the actual solution also has a continuous 2nd derivative.

Lemma 5.1.1 (Lemma of du Bois-Reymond). *Suppose $y \in \mathcal{U}$ is a local minimum of the functional $J : \mathcal{U} \to \mathbb{R}$ defined by*

$$J[y] = \int_a^b F(x, y, y') \, dx \qquad (5.1.6)$$

where F has continuous partial derivatives up to order 2 and

$$\frac{\partial^2 F}{\partial (y')^2} \neq 0, \qquad \forall x \in I$$

Then $y \in \mathcal{C}^2(I)$.

Proof. Consider the necessary condition on the 1st variation

$$\delta J = \int_a^b \left[\frac{\partial F}{\partial y} \delta y + \frac{\partial F}{\partial y'} \delta y' \right] dx = 0, \qquad \forall \delta y \in \mathcal{V} = \mathcal{C}_0^1(I)$$

We apply lemma 4.6.3 to the integral and conclude that

$$\frac{d}{dx} \frac{\partial F}{\partial y'} = \frac{\partial F}{\partial y}$$

Subtracting, we obtain the Euler equation

$$\frac{\partial F}{\partial y} - \frac{d}{dx} \frac{\partial F}{\partial y'} = 0$$

Lemma 4.6.3 also states that these functions are continuous on I. We explicitly compute the derivative $\frac{d}{dx} \frac{\partial F}{\partial y'}$ using chain rule.

$$\frac{d}{dx} \frac{\partial F}{\partial y'} = \frac{\partial^2 F}{\partial x \partial y'} + \frac{\partial^2 F}{\partial y \partial y'} y' + \frac{\partial^2 F}{\partial (y')^2} y''$$

Substituting into the Euler equation and solving for y'' gives the expression

$$y'' = \left[\frac{\partial F}{\partial y} - \frac{\partial^2 F}{\partial x \partial y'} - \frac{\partial^2 F}{\partial y \partial y'} y' \right] \left[\frac{\partial^2 F}{\partial (y')^2} \right]^{-1}$$

Every term on the right hand side is continuous and the denominator is non-zero. Thus, we have $y'' \in \mathcal{C}^2(I)$. $\square$

5.1. FIXED END POINTS

The derivation of the Euler equation contains some subtle points.

- If the Lemma of du Bois-Reymond is applicable, then the solution space $\mathcal{U}$ has increased order of continuity. (e.g. $y \in \mathcal{C}^2(I)$ instead of $y \in \mathcal{C}^1(I)$). The theorem can be extended to functionals containing higher order derivatives and multiple functions. However, we will not explore the issue any further.

- Increasing the order of continiuty of y to $\mathcal{C}^2(I)$ ensures that the expression

$$\frac{\partial F}{\partial y} - \frac{d}{dx}\frac{\partial F}{\partial y'} = \frac{\partial F}{\partial y} - \frac{\partial^2 F}{\partial x \partial y'} + \frac{\partial^2 F}{\partial y \partial y'}y' + \frac{\partial^2 F}{\partial (y')^2}y''$$

is continuous because F has continuous partials up to order 2 in each of its arguments. Similar issues are encountered in proofs later in this text. However, we choose not to explicitly increase the order of continuity of the solution space. The reason is that

$$\frac{\partial^2 F}{\partial (y')^2} = 0, \quad \forall x \in I$$

allows continuity of

$$\frac{\partial F}{\partial y} - \frac{d}{dx}\frac{\partial F}{\partial y'}$$

with only $y \in \mathcal{C}^1(I)$. We do not wish to exclude this possibility.

- Lemma 4.6.3 implies the Euler equation

$$\frac{\partial F}{\partial y} - \frac{d}{dx}\frac{\partial F}{\partial y'} = 0$$

Integration by parts is not required to prove this equation. Analogues to this theorem exist for other variational problems but are not discussed in this book.

In order to avoid any further difficulties, we henceforth assume that *differential expressions inside integrals are continuous!* The reader should be aware of this fact whenever the Fundamental Lemma 4.6.1 is used in this text.

This page intentionally left blank.

5.1. FIXED END POINTS

The 2nd variation is computed using property (4.4.5).

$$\delta^2 J = \delta^2 \int_a^b F\, dx = \int_a^b \delta^2 F\, dx$$

Inserting the variation $\delta^2 F$ into the integral gives

$$\delta^2 J = \frac{1}{2} \int_a^b \left[\frac{\partial^2 F}{\partial y^2}(\delta y)^2 + 2\frac{\partial^2 F}{\partial y \partial y'}\delta y\, \delta y' + \frac{\partial^2 F}{\partial (y')^2}(\delta y')^2 \right] dx \tag{5.1.7}$$

To investigate the sign of the 2nd variation, we transform the expression for $\delta^2 J$ to a more convenient form. Assume that F has continuous derivatives up to order 3 in each of its arguments. The middle term in the integral of equation (5.1.7) is integrated by parts. We first note that the anti-derivative of $\delta y\, \delta y'$.

$$\int \delta y\, \delta y'\, dx = \frac{1}{2}(\delta y)^2$$

This result is used when integrating by parts.

$$2\int_a^b \frac{\partial^2 F}{\partial y \partial y'}\delta y\, \delta y'\, dx = -\int_a^b \left[\frac{d}{dx}\frac{\partial^2 F}{\partial y \partial y'}\right](\delta y)^2 dx + \left[\frac{\partial^2 F}{\partial y \partial y'}(\delta y)^2\right]\Bigg|_a^b$$

The boundary terms vanish because $\delta y(a) = \delta y(b) = 0$.

$$2\int_a^b \frac{\partial^2 F}{\partial y \partial y'}\delta y\, \delta y'\, dx = -\int_a^b \left[\frac{d}{dx}\frac{\partial^2 F}{\partial y \partial y'}\right](\delta y)^2 dx$$

Equation (5.1.7) is transformed to

$$\delta^2 J = \frac{1}{2}\int_a^b \left[\left(\frac{\partial^2 F}{\partial y^2} - \frac{d}{dx}\frac{\partial^2 F}{\partial y \partial y'}\right)(\delta y)^2 + \frac{\partial^2 F}{\partial (y')^2}(\delta y')^2\right] dx$$

We write this result as

$$\delta^2 J = \int_a^b \left[P(x)(\delta y')^2 + Q(x)(\delta y)^2\right] dx \tag{5.1.8}$$

$$P(x) = \frac{1}{2}\frac{\partial^2 F}{\partial (y')^2} \tag{5.1.9}$$

$$Q(x) = \frac{1}{2}\left(\frac{\partial^2 F}{\partial y^2} - \frac{d}{dx}\frac{\partial^2 F}{\partial y \partial y'}\right) \tag{5.1.10}$$

Lemma 5.1.2. *Consider $\delta^2 J$ defined by equation (5.1.8). A necessary condition for $\delta^2 J \geq 0$ for all admissible variations $\delta y \in \mathcal{V}$ is that $P(x) \geq 0$ for $x \in I = [a,b]$.*

Proof. We repeat the proof by contradiction given by [9]. Assume that $\delta^2 J \geq 0$ for all admissible variations. Suppose $P(\xi) = -2\beta$ for some $\xi \in (a,b)$ and $\beta > 0$. The continuity of $P(x)$ ensures that there is a $\alpha > 0$ such that $P(x) < -\beta$ for $x \in (\xi - \alpha, \xi + \alpha)$. Consider the following admissible variation:

$$\delta y(x) = \begin{cases} \sin^2\left[\frac{\pi(x-\xi)}{\alpha}\right] & \text{if } x \in (\xi - \alpha, \xi + \alpha) \\ 0 & \text{otherwise} \end{cases}$$

with its derivative

$$\delta y'(x) = \begin{cases} \left(\frac{\pi}{\alpha}\right) \sin\left[\frac{2\pi(x-\xi)}{\alpha}\right] & \text{if } x \in (\xi - \alpha, \xi + \alpha) \\ 0 & \text{otherwise} \end{cases}$$

Clearly we have $\delta y \in \mathcal{V}$. We compute $\delta^2 J$ for this variation using using equation (5.1.8).

$$\begin{aligned}\delta^2 J &= \int_a^b \left[P(x)(\delta y')^2 + Q(x)(\delta y)^2\right] dx \\ &= \int_{\xi-\alpha}^{\xi+\alpha} P(x) \left(\frac{\pi}{\alpha}\right)^2 \sin^2\left[\frac{2\pi(x-\xi)}{\alpha}\right] dx \\ &+ \int_{\xi-\alpha}^{\xi+\alpha} Q(x) \sin^4\left[\frac{\pi(x-\xi)}{\alpha}\right]\end{aligned}$$

The integral terms require further analysis. The first integral obeys the following inequality:

$$\int_{\xi-\alpha}^{\xi+\alpha} P(x) \left(\frac{\pi}{\alpha}\right)^2 \sin^2\left[\frac{2\pi(x-\xi)}{\alpha}\right] dx < -\beta \left(\frac{\pi}{\alpha}\right)^2 \int_{\xi-\alpha}^{\xi+\alpha} \sin^2\left[\frac{2\pi(x-\xi)}{\alpha}\right] dx$$

because $P(x) < -\beta$ for $x \in (\xi - \alpha, \xi + \alpha)$. We now compute the integral on the right hand side. Let $\theta = x - \xi$ and $\omega = 2\pi/\alpha$.

5.1. FIXED END POINTS

$$\begin{aligned}
-\beta\left(\frac{\pi}{\alpha}\right)^2 \int_{\xi-\alpha}^{\xi+\alpha} \sin^2\left[\frac{2\pi(x-\xi)}{\alpha}\right] dx &= -\beta\left(\frac{\pi}{\alpha}\right)^2 \int_{-\alpha}^{\alpha} \sin^2(\omega\theta)\, d\theta \\
&= -2\beta\left(\frac{\pi}{\alpha}\right)^2 \int_0^{\alpha} \sin^2(\omega\theta)\, d\theta \\
&= -2\beta\left(\frac{\pi}{\alpha}\right)^2 \left[\frac{\theta}{2} - \frac{\sin(2\omega\theta)}{4\omega}\right]\Big|_0^{\alpha} \\
&= -2\beta\left(\frac{\pi}{\alpha}\right)^2 \left[\frac{\alpha}{2} - \frac{\sin(2\omega\alpha)}{4\omega}\right] \\
&= -2\beta\left(\frac{\pi}{\alpha}\right)^2 \left[\frac{\alpha}{2} - \frac{\alpha\sin(4\pi)}{8\pi}\right] \\
&= -\frac{\beta\pi^2}{\alpha}
\end{aligned}$$

The other integral satisfies

$$\int_{\xi-\alpha}^{\xi+\alpha} Q(x) \sin^4\left[\frac{\pi(x-\xi)}{\alpha}\right] < \|Q\|_{\sup} \int_{\xi-\alpha}^{\xi+\alpha} \sin^4\left[\frac{\pi(x-\xi)}{\alpha}\right]$$

We set $\omega = \pi/\alpha$ and observe that the integral is bounded.

$$\begin{aligned}
\|Q\|_{\sup} \int_{\xi-\alpha}^{\xi+\alpha} \sin^4\left[\frac{\pi(x-\xi)}{\alpha}\right] &= \|Q\|_{\sup} \int_{-\alpha}^{\alpha} \sin^4(\omega\theta)\, d\theta \\
&< \|Q\|_{\sup} \int_{-\alpha}^{\alpha} d\theta \\
&< 2\alpha \|Q\|_{\sup}
\end{aligned}$$

Combining these results gives

$$\delta^2 J < -\frac{\beta\pi^2}{\alpha} + 2\alpha \|Q\|_{\sup}$$

For sufficiently small α, we can force $\delta^2 J < 0$. This contradicts the hypothesis that $\delta^2 J \geq 0$ for all admissible variations. Hence the statement of the lemma is true. □

This lemma combined with the 2nd order necessary condition $\delta^2 J \geq 0$ proves the Legendre condition which is stated on the next page.

Theorem 5.1.2 (Legendre Condition). *Consider the closed interval $I = [a, b]$, the set of functions $\mathcal{U} = \{y \in \mathcal{C}^1(I) \mid y(a) = y_a \text{ and } y(b) = y_b\}$, and the functional $J : \mathcal{U} \to \mathbb{R}$ defined by*

$$J[y] = \int_a^b F(x, y, y') \, dx \tag{5.1.11}$$

where F has continuous partial derivatives up to order 3. A necessary condition for $y(x)$ to be a local minimum of J is that the Legendre condition

$$\frac{\partial^2 F}{\partial (y')^2} \geq 0 \tag{5.1.12}$$

be satisfied at every point of the curve.

We provide another derivation of the Euler equation (5.1.5) using the alternate variational formulation. Let $\delta y(x) = \varepsilon \, \eta(x)$ where $\eta \in \mathcal{V} = \mathcal{C}_0^1(I)$ and ε is a small parameter. Clearly $\delta y \in \mathcal{V}$. The first derivative of δy is computed as

$$\delta y'(x) = \varepsilon \, \eta'(x)$$

The 2nd order Taylor series expansion of the functional J is given by equation (4.3.2).

$$J[y + \varepsilon \eta] = J[y] + \varepsilon \, J_1[y][\eta] + \frac{1}{2} \varepsilon^2 \, J_2[y][\eta] + \mu \varepsilon^2 \, \|\eta\|_{1sup}^2 \tag{5.1.13}$$

$$J_1[y][\eta] = \int_a^b \left[\frac{\partial F}{\partial y} \eta + \frac{\partial F}{\partial y'} \eta' \right] dx \tag{5.1.14}$$

$$J_2[y][\eta] = \int_a^b \left[\frac{\partial^2 F}{\partial y^2} \eta^2 + 2 \frac{\partial^2 F}{\partial y \partial y'} \eta \eta' + \frac{\partial^2 F}{\partial (y')^2} (\eta')^2 \right] dx \tag{5.1.15}$$

where $\mu \to 0$ as $\varepsilon \to 0$. Notice that $\delta J = J_1$ and $\delta^2 J = \frac{1}{2} J_2$ according to the formal definitions of 1st and 2nd variation.

Recalling equation (4.3.3) to extract the 1st variation.

$$\frac{d}{d\varepsilon} J[y + \varepsilon \eta] \bigg|_{\varepsilon = 0} = J_1[y][\eta]$$

A necessary requirement for a local minimum is

$$\frac{d}{d\varepsilon} J[y + \varepsilon \eta] \bigg|_{\varepsilon = 0} = 0 \quad \Rightarrow \quad J_1[y][\eta] = 0 \quad \Rightarrow \quad \delta J = 0 \tag{5.1.16}$$

5.1. FIXED END POINTS

The 1st order necessary condition is independent of the choice of η. We Integrate equation (5.1.14) by parts and use the fact that η vanishes at the boundary.

$$\begin{aligned} J_1[y][\eta] &= \int_a^b \left[\frac{\partial F}{\partial y} - \frac{d}{dx}\frac{\partial F}{\partial y'} \right] \eta\, dx + \left[\frac{\partial F}{\partial y'} \eta \right]_a^b \\ &= \int_a^b \left[\frac{\partial F}{\partial y} - \frac{d}{dx}\frac{\partial F}{\partial y'} \right] \eta\, dx \end{aligned}$$

The requirement $J_1[y][\eta] = 0$ for all $\eta \in \mathcal{V}$ and application of the Fundamental Lemma 4.6.1 gives the Euler equation.

$$\frac{\partial F}{\partial y} - \frac{d}{dx}\frac{\partial F}{\partial y'} = 0 \qquad (5.1.17)$$

The 2nd order necessary condition is provided by using equation (4.3.4).

$$\left. \frac{d^2}{d\varepsilon^2} J[y+\varepsilon\eta] \right|_{\varepsilon=0} = J_2[y][\eta]$$

Thus, we have the condition

$$\left. \frac{d^2}{d\varepsilon^2} J[y+\varepsilon\eta] \right|_{\varepsilon=0} \geq 0 \quad \Rightarrow \quad J_2[y][\eta] \geq 0 \quad \Rightarrow \delta^2 J \geq 0 \qquad (5.1.18)$$

These alternate procedures to derive the Euler equation were provided for comparison to other texts. However, such procedures do not provide any additional insight and will not be discussed further.

5.2 Special Cases

5.2.1 $F \equiv F(y, y')$

The Euler equation (5.1.5) is

$$\frac{\partial F}{\partial y} - \frac{d}{dx}\frac{\partial F}{\partial y'} = 0 \qquad (5.2.1)$$

If F does not depend on x, then $F \equiv F(y, y')$. We compute the derivative d/dx with chain rule.

$$\frac{\partial F}{\partial y} - \frac{\partial^2 F}{\partial y \partial y'}y' - \frac{\partial^2 F}{\partial (y')^2}y'' = 0$$

Multiplying both sides by y' gives

$$\frac{\partial F}{\partial y}y' - \frac{\partial^2 F}{\partial y \partial y'}(y')^2 - \frac{\partial^2 F}{\partial (y')^2}y'y'' = 0$$

Some additional algebra and calculus provides

$$\left[\frac{\partial F}{\partial y}y' + \frac{\partial F}{\partial y'}y''\right] - \left[\frac{\partial F}{\partial y'}y'' + \frac{\partial^2 F}{\partial y \partial y'}(y')^2 + \frac{\partial^2 F}{\partial (y')^2}y'y''\right] = 0$$

$$\frac{dF}{dx} - \left[\frac{\partial F}{\partial y'}\frac{d}{dx}(y') + y'\frac{d}{dx}\frac{\partial F}{\partial y'}\right] = 0$$

$$\frac{dF}{dx} - \frac{d}{dx}\left[y'\frac{\partial F}{\partial y'}\right] = 0$$

$$\frac{d}{dx}\left[F - y'\frac{\partial F}{\partial y'}\right] = 0$$

This implies that

$$F - y'\frac{\partial F}{\partial y'} = C \qquad (5.2.2)$$

where C is a constant. This 1st order differential equation is solved to obtain y.

5.2.2 $F \equiv F(x, y')$

If F does not depend on y, then $F \equiv F(x, y')$. The Euler equation (5.2.1) becomes

$$\frac{d}{dx}\frac{\partial F}{\partial y'} = 0$$

This implies that

$$\frac{\partial F}{\partial y'} = C \tag{5.2.3}$$

where C is a constant. If we can solve for y', we get the differential equation $y' = f(x, C)$ because F does not contain y. This differential equation is solved by integration.

5.2.3 $F \equiv F(x, y)$

If F does not depend on y', then $F \equiv F(x, y)$. The Euler equation (5.2.1) becomes

$$\frac{\partial F}{\partial y} = 0 \tag{5.2.4}$$

There is no differential equation to solve. Equation (5.2.4) provides an algebraic condition and y is given implicitly as a function of x.

5.3 Shortest Distance Between Two Points

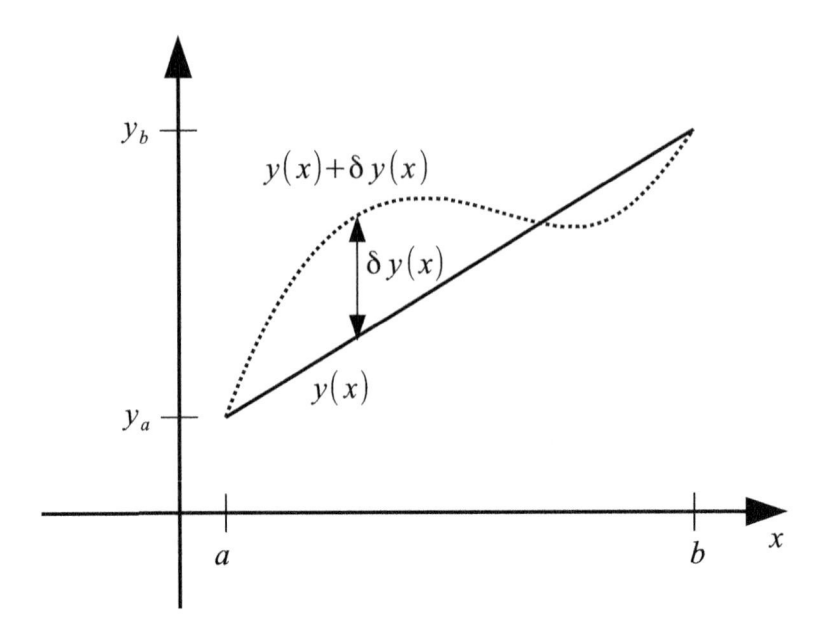

Figure 5.3.1: Shortest distance between two points

Consider the obvious statement that the shortest distance between two points in the plane is a straight line. The calculus of variations can be used to prove this statement. We consider all possible smooth curves connecting the points and show that the curve with the smallest arc length is a line. We shall only consider non-vertical lines. The end points are (a, y_a) and (b, y_b) with $a < b$. The functional to minimize is

$$J[y] = \int_a^b \sqrt{1 + (y')^2}\, dx \tag{5.3.1}$$

$J : \mathcal{U} \to \mathbb{R}$ where $\mathcal{U} = \{y \in \mathcal{C}^1(I) \,|\, y(a) = y_a \text{ and } y(b) = y_b\}$. The endpoints are fixed and the problem is solved using the Euler differential equation with the function $F = [1 + (y')^2]^{1/2}$.

5.3. SHORTEST DISTANCE BETWEEN TWO POINTS

Since F does not contain y, we can use equation (5.2.3) to solve the problem.

$$\frac{\partial F}{\partial y'} = C$$

$$\frac{y'}{\sqrt{1+(y')^2}} = C$$

where C is a constant. We observe that $C \in (-1, 1)$. We show that y' must also be constant. Solving the equation for $(y')^2$ results in

$$\begin{aligned} y' &= C\sqrt{1+(y')^2} \\ (y')^2 &= C^2[1+(y')^2] \\ (1-C^2)(y')^2 &= C^2 \\ (y')^2 &= \frac{C^2}{1-C^2} \end{aligned}$$

The left hand side is a non-negative constant and so $(y')^2$ is constant. The continuity of y' means that it cannot change sign. In other words, it cannot jump between a positive and negative constant of the same magnitude. Hence, y' must be constant. Set $y'(x) = m$ and integrate to get $y(x) = mx + B$ where B is a constant. To satisfy the endpoint values $y(a) = y_a$ and $y(b) = y_b$, we solve

$$\begin{aligned} ma + B &= y_a \\ mb + B &= y_b \end{aligned}$$

for C and B. The solution is

$$m = \frac{y_b - y_a}{b - a}, \qquad B = y_a - ma$$

The equation of the line can be written in the familiar point-slope form

$$\begin{aligned} y(x) &= mx + B \\ &= mx + y_a - ma \\ &= m(x - a) + y_a \end{aligned}$$

5.4 Isoparametric Conditions

Some applications require the minimization of the functional

$$J[y] = \int_a^b F(x, y, y') \, dx \tag{5.4.1}$$

subject to $y(a) = y_a$, $y(b) = y_b$, and the condition

$$K[y] = \int_a^b G(x, y, y') \, dx = C \tag{5.4.2}$$

where C is a constant. F and G have continuous partial derivatives up to order 2. Set $I = [a, b]$. The set of possible solutions is

$$\mathcal{U} = \left\{ y \in \mathcal{C}^1(I) \mid y(a) = y_a \text{ and } y(b) = y_b \right\}$$

and the admissible variations are $\mathcal{V} = \mathcal{C}_0^1(I)$.

We use the method of Lagrange multipliers to solve the minimization problem. The variation of the functional K is zero because the variation of the constant C is zero. Let $G \equiv G(x, y, y')$.

$$\delta K = \delta \int_a^b G \, dx = \int_a^b \delta G \, dx = 0$$

We multiply δK by an unspecified constant λ and add the result to the variation δJ. In order words, we add zero to δJ. Integration by parts yields

$$\begin{aligned}
\delta J &= \delta J + \lambda \, \delta K \\
&= \int_a^b \delta F \, dx + \lambda \int_a^b \delta G \, dx \\
&= \int_a^b \left[\frac{\partial F}{\partial y} \delta y + \frac{\partial F}{\partial y'} \delta y' \right] dx + \lambda \int_a^b \left[\frac{\partial G}{\partial y} \delta y + \frac{\partial G}{\partial y'} \delta y' \right] dx \\
&= \int_a^b \left[\frac{\partial F}{\partial y} - \frac{d}{dx} \frac{\partial F}{\partial y'} \right] \delta y \, dx + \left[\frac{\partial F}{\partial y'} \delta y \right]\bigg|_a^b \\
&\quad + \lambda \int_a^b \left[\frac{\partial G}{\partial y} - \frac{d}{dx} \frac{\partial G}{\partial y'} \right] \delta y \, dx + \left[\lambda \frac{\partial G}{\partial y'} \delta y \right]\bigg|_a^b
\end{aligned}$$

The boundary terms vanish because $\delta y(a) = \delta y(b) = 0$. Combining the remaining integrals gives

$$\delta J = \int_a^b \left[\frac{\partial F}{\partial y} - \frac{d}{dx} \frac{\partial F}{\partial y'} + \lambda \left(\frac{\partial G}{\partial y} - \frac{d}{dx} \frac{\partial G}{\partial y'} \right) \right] \delta y \, dx$$

5.4. ISOPARAMETRIC CONDITIONS

The 1st order necessary condition requires $\delta J = 0$ for all admissible $\delta y \in \mathcal{V}$. Application of the Fundamental Lemma 4.6.1 means that the term in brackets must vanish.

$$\frac{\partial F}{\partial y} - \frac{d}{dx}\frac{\partial F}{\partial y'} + \lambda \left(\frac{\partial G}{\partial y} - \frac{d}{dx}\frac{\partial G}{\partial y'}\right) = 0$$

Theorem 5.4.1. *Consider the closed interval $I = [a, b]$, the set of functions $\mathcal{U} = \{y \in \mathcal{C}^1(I) \mid y(a) = y_a \text{ and } y(b) = y_b\}$, and the functional $J : \mathcal{U} \to \mathbb{R}$ defined by*

$$J[y] = \int_a^b F(x, y, y')\, dx \qquad (5.4.3)$$

where F has continuous partial derivatives up to order 2. A necessary condition for a local minimum of J over $\mathcal{U}$ subject to the isoparametric condition

$$K[y] = \int_a^b G(x, y, y')\, dx = C \qquad (5.4.4)$$

is that the following differential equation be satisfied:

$$\frac{\partial F}{\partial y} - \frac{d}{dx}\frac{\partial F}{\partial y'} + \lambda \left(\frac{\partial G}{\partial y} - \frac{d}{dx}\frac{\partial G}{\partial y'}\right) = 0 \qquad (5.4.5)$$

The solution for λ is obtained while solving equation (5.4.5) in conjunction with condition (5.4.4). Equation (5.4.5) is equivalent to the Euler differential equation for the augmented functional

$$\tilde{J}[y] = J[y] + \lambda K[y] = \int_a^b \tilde{F}(x, y, y', \lambda)\, dx \qquad (5.4.6)$$

$$\tilde{F}(x, y, y', \lambda) = F(x, y, y') + \lambda G(x, y, y') \qquad (5.4.7)$$

$$\frac{\partial \tilde{F}}{\partial y} - \frac{d}{dx}\frac{\partial \tilde{F}}{\partial y'} = 0 \qquad (5.4.8)$$

5.5 Shape of a Hanging Cable

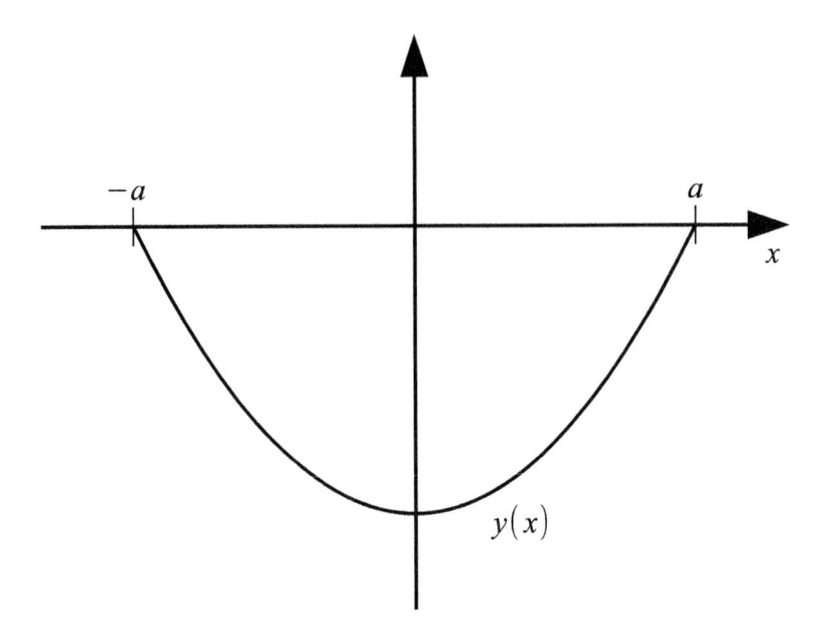

Figure 5.5.1: Hanging cable from two fixed points

An example of an isometric variational problem is provided by determining the shape of a hanging cable. Suppose that a cable of length L is suspended between two points having the same height. Figure 5.5.1 shows the situation. We set $y(-a) = y(a) = 0$ and require $L \geq 2a$. The shape of the cable is the function $y(x)$ which minimizes potential energy U. Consider a small segment of the cable with differential length ds. Let γ be the specific weight of the cable (i.e. weight per unit length). The weight of this segment is $\gamma \, ds$. The potential energy of the segment is computed as $\gamma y \, ds$. Recalling the formula for arc length, the total potential energy is

$$U[y] = \gamma \int_{-a}^{a} y\sqrt{1 + (y')^2}\, dx \tag{5.5.1}$$

This function is minimized subject to the total arc length condition

$$S[y] = \int_{-a}^{a} \sqrt{1 + (y')^2}\, dx = L \tag{5.5.2}$$

5.5. SHAPE OF A HANGING CABLE

The constant γ does not affect the minimization of the functional U. We write the augmented functional

$$\tilde{J}[y] = \frac{1}{\gamma}U[y] + \lambda S[y] = \int_{-a}^{a} (y + \lambda)\sqrt{1 + (y')^2}\, dx \qquad (5.5.3)$$

Let $\tilde{F} = (y + \lambda)\sqrt{1 + (y')^2}$ and notice that $\tilde{F}$ does not depend on x. We can use equation (5.2.2) to solve the resulting Euler equation. Let C be a constant.

$$\tilde{F} - y'\frac{\partial \tilde{F}}{\partial y'} = C$$

$$(y + \lambda)\sqrt{1 + (y')^2} - y'(y + \lambda)\frac{y'}{\sqrt{1 + (y')^2}} = C$$

$$(y + \lambda)\left[\sqrt{1 + (y')^2} - \frac{y'^2}{\sqrt{1 + (y')^2}}\right] = C$$

$$\frac{y + \lambda}{\sqrt{1 + (y')^2}} = C$$

$$y + \lambda = C\sqrt{1 + (y')^2}$$
$$(y + \lambda)^2 = C^2[1 + (y')^2]$$
$$(y + \lambda)^2 - C^2 = C^2(y')^2$$
$$\pm\sqrt{(y + \lambda)^2 - C^2} = Cy'$$

This differential equation is integrated by separation of variables.

$$\pm\frac{y'}{\sqrt{(y + \lambda)^2 - C^2}} = \frac{1}{C}$$

$$\pm\int\frac{dy}{\sqrt{(y + \lambda)^2 - C^2}} = \frac{1}{C}\int dx$$

$$\pm\cosh^{-1}\left(\frac{y + \lambda}{C}\right) = \frac{x}{C} + D$$

where D is a constant of integration. The $\pm$ gives the two branches of hyperbolic cosine. Any difficulties with the sign of C also resolved with the $\pm$. Thus, we can write

$$y = C\cosh\left(\frac{x}{C} + D\right) - \lambda \qquad (5.5.4)$$

The constants λ, C, and D are determined numerically using $y(-a) = y(a) = 0$ and equation (5.5.2). The most that we can say is that $D = 0$ due to symmetry of the problem.

5.6 Variable End Points

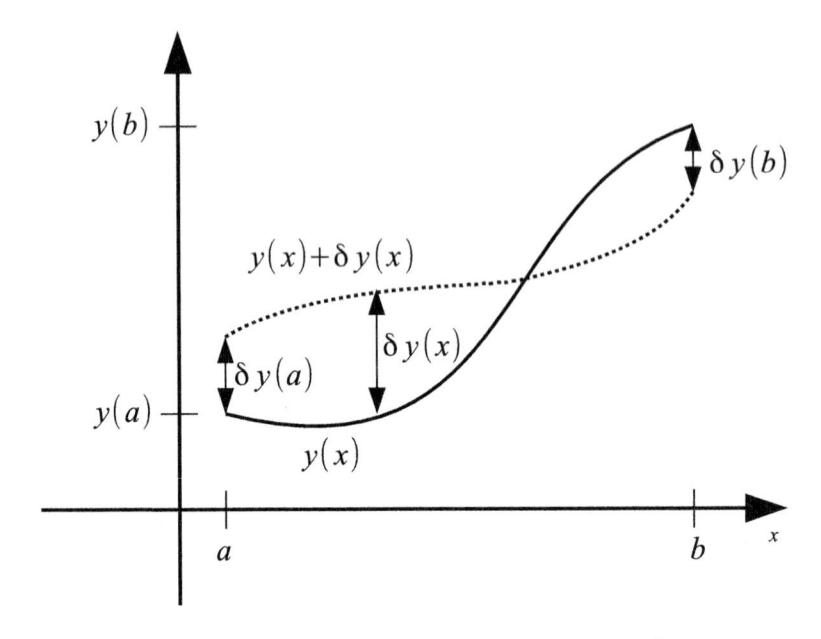

Figure 5.6.1: Variation $\delta y(x)$ of function $y(x)$

In this section we consider the more general functional

$$J[y] = \int_a^b F(x, y, y')\, dx + f(y(b)) - g(y(a)) \qquad (5.6.1)$$

where F has continuous derivatives up to order 2 in each of its arguments and $f, g \in \mathcal{C}^1(\mathbb{R})$. Unlike the previous section, the values of $y(x)$ are not fixed at the endpoints. Hence, the variation $\delta y(x)$ is also not required to vanish at the endpoints. Figure 5.6.1 demonstrates this situation. The set of possible solutions and admissible variations are now identical: $\mathcal{U} = \mathcal{V} = \mathcal{C}^1(I)$ where $I = [a, b]$. The techniques of calculus of variations are used to solve

$$y^* = \arg \operatorname*{local\,min}_{y \in \mathcal{U}} J[y] \qquad (5.6.2)$$

5.6. VARIABLE END POINTS

Two useful properties to compute the variation are: variation is a linear operator and variation commutes with integration. Notice that f and g are ordinary functions. We compute δJ in the following steps:

$$\begin{aligned}
\delta J &= \delta \int_a^b F\,dx + \delta[f(y(b))] - \delta[g(y(a))] \\
&= \int_a^b \delta F\,dx + f'(y(b))\,\delta y(b) - g'(y(a))\,\delta y(a) \\
&= \int_a^b \left[\frac{\partial F}{\partial y}\delta y + \frac{\partial F}{\partial y'}\delta y'\right] dx + f'(y(b))\,\delta y(b) - g'(y(a))\,\delta y(a) \\
&= \int_a^b \left[\frac{\partial F}{\partial y} - \frac{d}{dx}\frac{\partial F}{\partial y'}\right]\delta y\,dx + \left[\frac{\partial F}{\partial y'}\delta y\right]\Big|_a^b \\
&\quad + f'(y(b))\,\delta y(b) - g'(y(a))\,\delta y(a)
\end{aligned}$$

Integration by parts was used to produce the last step. We write the variation as

$$\delta J = \delta J_1 + \phi(b) - \psi(a) \tag{5.6.3}$$

$$\begin{aligned}
\delta J_1 &= \int_a^b \left[\frac{\partial F}{\partial y} - \frac{d}{dx}\frac{\partial F}{\partial y'}\right]\delta y\,dx \\
\phi(b) &= \left[\left(\frac{\partial F}{\partial y'} + f'\right)\delta y\right]\bigg|_b \\
\psi(a) &= \left[\left(\frac{\partial F}{\partial y'} + g'\right)\delta y\right]\bigg|_a
\end{aligned}$$

The vertical bar notation with b for $\phi(b)$ means that $\partial F/\partial y'$ is evaluated at $x = b$, $y = y(b)$, $y' = y'(b)$ and f' is evaluated at $y = y(b)$. In the same fashion, we have $\psi(a)$ implies $\partial F/\partial y'$ is evaluated at $x = a$, $y = y(a)$, $y' = y'(a)$ and g' is evaluated at $y = y(a)$.

The 1st order necessary condition $\delta J = 0$ will provide the necessary equations for a local minimum of J. If we choose $\delta y \in \mathcal{C}_0^1(I) \subset \mathcal{C}^1(I) = \mathcal{U}$, then $\phi(b) = \psi(a) = 0$ and $\delta J_1 = 0$. For this case, the Fundamental Lemma 4.6.1 states the term in brackets in δJ_1 must vanish. Hence, the Euler differential equation (5.1.5) must be satisfied. We now have that $\delta J_1 = 0$ for any δy. The only way $\phi(b)$ and $\psi(a)$ can vanish is if the natural boundary conditions are satisfied.

$$\left[\frac{\partial F}{\partial y'} + f'\right]\bigg|_b = 0, \qquad \left[\frac{\partial F}{\partial y'} + g'\right]\bigg|_a = 0$$

We have proved the theorem which appears on the next page.

Theorem 5.6.1. *Consider the closed interval* $I = [a, b]$, *the set of functions* $\mathcal{U} = \mathcal{C}^1(I)$, *and the functional* $J : \mathcal{U} \to \mathbb{R}$ *defined by*

$$J[y] = \int_a^b F(x, y, y') \, dx + f(y(b)) - g(y(a)) \tag{5.6.4}$$

where F *has continuous partial derivatives up to order 2 and* $f, g \in \mathcal{C}^1(\mathbb{R})$. *A necessary condition for a local minimum of* J *over* $\mathcal{U}$ *is that the following equations be satisfied:*

$$\frac{\partial F}{\partial y} - \frac{d}{dx}\frac{\partial F}{\partial y'} = 0 \tag{5.6.5}$$

$$\left[\frac{\partial F}{\partial y'} + f'\right]\bigg|_b = 0, \quad \left[\frac{\partial F}{\partial y'} + g'\right]\bigg|_a = 0 \tag{5.6.6}$$

We can compute the 2nd variation, assuming that $f, g \in \mathcal{C}^2(\mathbb{R})$, as follows:

$$\delta^2 J = \int_a^b \delta^2 F \, dx + \delta^2[f(y(b))] - \delta^2[g(y(a))]$$

The resulting equation is

$$\begin{aligned}\delta^2 J &= \frac{1}{2}\int_a^b \left[\frac{\partial^2 F}{\partial y^2}(\delta y)^2 + 2\frac{\partial^2 F}{\partial y \partial y'}\delta y \, \delta y' + \frac{\partial^2 F}{\partial (y')^2}(\delta y')^2\right] dx \\ &\quad + f''(y(b))(\delta y)^2 - g''(y(a))(\delta y)^2\end{aligned} \tag{5.6.7}$$

This page intentionally left blank.

5.7 Elastic Bar

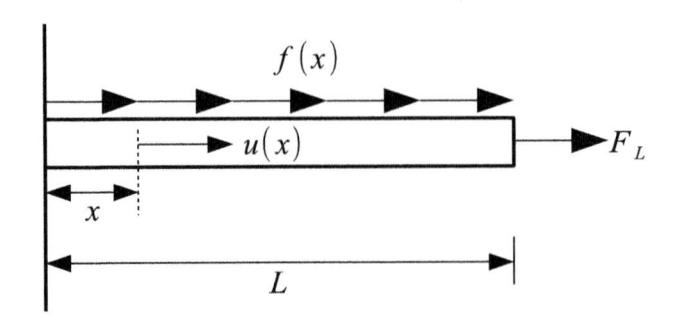

Figure 5.7.1: Elastic bar

A prismatic elastic bar is shown in figure 5.7.1. The original length of the bar is L and its cross-sectional area is A. The bar is loaded tangentially along its length by a distributed load $f(x)$. It is also loaded at the right end by a load F_L. The bar is restrained from motion at the left end. $u(x)$ is the displacement at location x. The corresponding strain is $u'(x)$.

The equilibrium displacement $u(x)$ of the bar minimizes the total potential energy Π.

$$\Pi[u] = U(u') - V(u)$$
$$U(u') = \frac{1}{2}EA \int_0^L [u'(x)]^2 dx$$
$$V(u) = \int_0^L f(x)\,u(x)\,dx + F_L\,u(L)$$

where U is the strain energy and V is the work done by external forces. E is the modulus of elasticity and is a positive constant. The functional can also be written in the form

$$\Pi[u] = \int_0^L \left\{ \frac{1}{2}EA[u'(x)]^2 - f(x)\,u(x) \right\} dx - F_L\,u(L) \tag{5.7.1}$$

We shall assume that the distributed load is continuous: $f \in \mathcal{C}([0,L])$. The continuity of the material requires that the displacement and strain both be continuous. Furthermore, we have the boundary condition $u(0) = 0$. The solution space is formulated as

$$\mathcal{U} = \{ u \in \mathcal{C}^1([0,L]) \mid u(0) = 0 \}$$

No variation is allowed where the displacement is prescribed. The variational space is defined as: $\mathcal{V} = \mathcal{U}$.

5.7. ELASTIC BAR

The results of the previous section are used to give necessary conditions for a minimum. Write the functional Π as

$$\Pi[u] = \int_0^L \tilde{F}(u, u') \, dx + \phi(u(L))$$

$$\tilde{F}(u, u') = \frac{1}{2} EA(u')^2 - fu$$

$$\phi(u) = -F_L u$$

The necessary differential equation is

$$\frac{\partial \tilde{F}}{\partial u} - \frac{d}{dx} \frac{\partial \tilde{F}}{\partial u'} = 0$$

$$-f - EA u'' = 0$$

We also must satisfy the natural boundary condition

$$\left[\frac{\partial \tilde{F}}{\partial u'} + \phi' \right]_L = 0$$

$$EA u'(L) - F_L = 0$$

Thus we have

$$-EA u''(x) = f(x) \tag{5.7.2}$$

with the boundary conditions

$$u(0) = 0, \qquad EA u'(L) = F_L \tag{5.7.3}$$

The differential equation (5.7.2) can be also derived from equilibrium considerations combined with Hooke's law which relates stress to strain: $\sigma(x) = E u'(x)$. Further information on these topics can be found in [13, 14].

The 2nd variation of the total potential energy is easily computed.

$$\delta^2 \Pi = \frac{1}{2} \int_0^L \left[\frac{\partial^2 \tilde{F}}{\partial u^2} (\delta u)^2 + 2 \frac{\partial^2 \tilde{F}}{\partial u \partial u'} \delta u \, \delta u' + \frac{\partial^2 \tilde{F}}{\partial (u')^2} (\delta u')^2 \right] dx$$

$$= \frac{1}{2} \int_0^L \left[EA(\delta u')^2 \right] dx$$

$$= \frac{1}{2} EA \int_0^L (\delta u')^2 dx$$

We can show that the 2nd variation is strictly positive for all admissible non-zero variations. Clearly we have

$$\delta^2 \Pi = \frac{1}{2} EA \int_0^L (\delta u')^2 dx \geq 0$$

The only way $\delta^2\Pi = 0$ is when $\delta u'(x) = 0$ for all $x \in [0, L]$ because $\delta u \in \mathcal{V} \subset \mathcal{C}^1([a,b])$. If this is the case, then $\delta u(x) = c$ for some constant c. However, $\delta u(0) = 0$ implies $c = 0$. Thus for all non-zero variations, $\delta^2\Pi > 0$ and this confirms a local minimum. Furthermore, it is a global minimum because $\delta^2\Pi > 0$ does not depend on u.

The case where $\delta u'(x) = 0$ for all $x \in [0, L]$ has physical significance. Since the solutions and variations come from the same space, we must allow the solution to also be an admissible variation. Hence, we can set $\delta u(x) = u(x)$. The solution $u(x) = c$ is a rigid body translation of the bar. Rigid body motions do not properly restrain the system. Also, they violate the uniqueness of the solution and must be excluded.

Chapter 6

Function with 1st and 2nd Derivatives

6.1 Fixed End Points

Consider the functional which contains $y(x)$ and it first two derivatives

$$J[y] = \int_a^b F(x, y, y', y'') \, dx \qquad (6.1.1)$$

where F has continuous derivatives up to order 3 in each of its arguments. The function $y(x)$ and its derivative $y'(x)$ have fixed values at the endpoints. The values of the derivatives at the endpoints are needed to determine a unique solution. Let $I = [a, b]$ be the domain of $y(x)$ and define

$$\mathcal{U} = \left\{ y \in \mathcal{C}^2(I) \,|\, y(a) = y_a, \, y(b) = y_b, \, y'(a) = y'_a, \, y'(b) = y'_b \right\}$$

as the set of possible solutions to

$$y^* = \arg \operatorname*{local\,min}_{y \in \mathcal{U}} J[y] \qquad (6.1.2)$$

The fact that F contains the 2nd derivative of y requires an increase in the order of continuity of the derivatives of y. This is captured in the new definition of $\mathcal{U}$.

The set $\mathcal{V}$ of admissible variations must also be defined. The variation δy must have at least the same order of continuous differentiability as y. This implies $\mathcal{V} \subset \mathcal{C}^2(I)$. Variations and their derivatives must vanish at the endpoints to ensure that $y + \delta y \in \mathcal{U}$. Hence, the set of admissible variations is

$$\mathcal{V} = \left\{ \delta y \in \mathcal{C}^2(I) \,|\, \delta y(a) = \delta y(b) = \delta y'(a) = \delta y'(b) = 0 \right\}$$

CHAPTER 6. FUNCTION WITH 1ST AND 2ND DERIVATIVES

The 1st variation of J was already provided in a previous chapter. However, we again use the simplified procedure to rederive the equations. Let $F \equiv F(x, y, y', y'')$. We compute δF and use integration by parts multiple times.

$$\begin{aligned}
\delta J &= \int_a^b \delta F \, dx \\
&= \int_a^b \left[\frac{\partial F}{\partial y} \delta y + \frac{\partial F}{\partial y'} \delta y' + \frac{\partial F}{\partial y''} \delta y'' \right] dx \\
&= \int_a^b \frac{\partial F}{\partial y} \delta y \, dx + \int_a^b \frac{\partial F}{\partial y'} \delta y' \, dx + \int_a^b \frac{\partial F}{\partial y''} \delta y'' \, dx \\
&= \int_a^b \frac{\partial F}{\partial y} \delta y \, dx - \int_a^b \left[\frac{d}{dx} \frac{\partial F}{\partial y'} \right] \delta y \, dx + \left[\frac{\partial F}{\partial y'} \delta y \right]\Big|_a^b \\
&\quad - \int_a^b \left[\frac{d}{dx} \frac{\partial F}{\partial y''} \right] \delta y' \, dx + \left[\frac{\partial F}{\partial y''} \delta y' \right]\Big|_a^b \\
&= \int_a^b \frac{\partial F}{\partial y} \delta y \, dx - \int_a^b \left[\frac{d}{dx} \frac{\partial F}{\partial y'} \right] \delta y \, dx + \left[\frac{\partial F}{\partial y'} \delta y \right]\Big|_a^b \\
&\quad + \int_a^b \left[\frac{d^2}{dx^2} \frac{\partial F}{\partial y''} \right] \delta y \, dx - \left[\frac{d}{dx} \frac{\partial F}{\partial y''} \delta y \right]\Big|_a^b + \left[\frac{\partial F}{\partial y''} \delta y' \right]\Big|_a^b
\end{aligned}$$

Recombining the integrals gives

$$\begin{aligned}
\delta J &= \int_a^b \left[\frac{\partial F}{\partial y} - \frac{d}{dx} \frac{\partial F}{\partial y'} + \frac{d^2}{dx^2} \frac{\partial F}{\partial y''} \right] \delta y \, dx \\
&\quad + \left[\frac{\partial F}{\partial y'} \delta y \right]\Big|_a^b - \left[\frac{d}{dx} \frac{\partial F}{\partial y''} \delta y \right]\Big|_a^b + \left[\frac{\partial F}{\partial y''} \delta y' \right]\Big|_a^b
\end{aligned}$$

The boundary terms vanish and we have

$$\delta J = \int_a^b \left[\frac{\partial F}{\partial y} - \frac{d}{dx} \frac{\partial F}{\partial y'} + \frac{d^2}{dx^2} \frac{\partial F}{\partial y''} \right] \delta y \, dx$$

Application of the Fundamental Lemma 4.6.1 to the 1st order necessary condition $\delta J = 0$ implies

$$\frac{\partial F}{\partial y} - \frac{d}{dx} \frac{\partial F}{\partial y'} + \frac{d^2}{dx^2} \frac{\partial F}{\partial y''} = 0$$

The solution to this differential equation is a necessary condition for a minimum of the functional J.

6.1. FIXED END POINTS

Another derivation is provided by applying lemma 4.6.5 to

$$\delta J = \int_a^b \left[\frac{\partial F}{\partial y} \delta y + \frac{\partial F}{\partial y'} \delta y' + \frac{\partial F}{\partial y''} \delta y'' \right] dx = 0$$

to obtain

$$\frac{\partial F}{\partial y} - \frac{d}{dx} \frac{\partial F}{\partial y'} + \frac{d^2}{dx^2} \frac{\partial F}{\partial y''} = 0$$

These results are summarized in the next theorem.

Theorem 6.1.1. *Consider the closed interval $I = [a, b]$, the set of functions*

$$\mathcal{U} = \{ y \in C^2(I) \mid y(a) = y_a,\ y(b) = y_b,\ y'(a) = y'_a,\ y'(b) = y'_b \}$$

and the functional $J : \mathcal{U} \to \mathbb{R}$ defined by

$$J[y] = \int_a^b F(x, y, y', y'')\, dx \tag{6.1.3}$$

where F has continuous partial derivatives up to order 3. A necessary condition for a local minimum of J over $\mathcal{U}$ is that the following differential equation be satisfied:

$$\frac{\partial F}{\partial y} - \frac{d}{dx} \frac{\partial F}{\partial y'} + \frac{d^2}{dx^2} \frac{\partial F}{\partial y''} = 0 \tag{6.1.4}$$

We will not extensively use the 2nd variation, but provide it for the sake of completeness. The 2nd variation is computed using

$$\delta^2 J = \int_a^b \delta^2 F\, dx$$

Inserting the expression for $\delta^2 F$, we get

$$\begin{aligned}
\delta^2 J &= \frac{1}{2} \int_a^b \left[\frac{\partial^2 F}{\partial y^2} (\delta y)^2 + \frac{\partial^2 F}{\partial (y')^2} (\delta y')^2 + \frac{\partial^2 F}{\partial (y'')^2} (\delta y'')^2 \right] dx \\
&\quad + \int_a^b \left[\frac{\partial^2 F}{\partial y \partial y'} \delta y\, \delta y' + \frac{\partial^2 F}{\partial y' \partial y''} \delta y'\, \delta y'' + \frac{\partial^2 F}{\partial y \partial y''} \delta y\, \delta y'' \right] dx \quad (6.1.5)
\end{aligned}$$

6.2 Variable End Points

Consider the more general functional

$$J[y] = \int_a^b F(x, y, y', y'') \, dx + f(y(b)) - g(y(a)) + h(y'(b)) - p(y'(a)) \quad (6.2.1)$$

where F has continuous partial derivatives up to order 3 in each of its arguments and $f, g, h, p \in \mathcal{C}^1(\mathbb{R})$. The values of $y(x)$ are not fixed at the endpoints and thus the variation $\delta y(x)$ is not required to vanish at the endpoints. Let $\mathcal{U} = \mathcal{V} = \mathcal{C}^2(I)$ with $I = [a, b]$. The minimization problem is

$$y^* = \arg \operatorname*{local\,min}_{y \in \mathcal{U}} J[y] \quad (6.2.2)$$

The 1st variation is computed using our previous knowledge.

$$\begin{aligned}
\delta J &= \delta \int_a^b F \, dx + \delta[f(y(b))] - \delta[g(y(a))] + \delta[h(y'(b))] - \delta[p(y'(a))] \\
&= \int_a^b \delta F \, dx + f'(y(b)) \, \delta y(b) - g'(y(a)) \, \delta y(a) \\
&\quad + h'(y'(b)) \, \delta y'(b) - p'(y'(a)) \, \delta y'(a) \\
&= \int_a^b \left[\frac{\partial F}{\partial y} \delta y + \frac{\partial F}{\partial y'} \delta y' + \frac{\partial F}{\partial y''} \delta y'' \right] dx \\
&\quad + f'(y(b)) \, \delta y(b) - g'(y(a)) \, \delta y(a) \\
&\quad + h'(y'(b)) \, \delta y'(b) - p'(y'(a)) \, \delta y'(a) \\
&= \int_a^b \left[\frac{\partial F}{\partial y} - \frac{d}{dx}\frac{\partial F}{\partial y'} + \frac{d^2}{dx^2}\frac{\partial F}{\partial y''} \right] \delta y \, dx \\
&\quad + \left[\frac{\partial F}{\partial y'} \delta y \right]_a^b - \left[\frac{d}{dx}\frac{\partial F}{\partial y''} \delta y \right]_a^b + \left[\frac{\partial F}{\partial y''} \delta y' \right]_a^b \\
&\quad + f'(y(b)) \, \delta y(b) - g'(y(a)) \, \delta y(a) \\
&\quad + h'(y'(b)) \, \delta y'(b) - p'(y'(a)) \, \delta y'(a)
\end{aligned}$$

Collecting like terms, the 1st variation is written as

$$\delta J = \delta J_1 + \phi(b) - \psi(a) + \gamma(b) - \mu(a) \quad (6.2.3)$$

$$\delta J_1 = \int_a^b \left[\frac{\partial F}{\partial y} - \frac{d}{dx}\frac{\partial F}{\partial y'} + \frac{d^2}{dx^2}\frac{\partial F}{\partial y''} \right] \delta y \, dx$$

6.2. VARIABLE END POINTS

$$\phi(b) = \left[\left(\frac{\partial F}{\partial y'} - \frac{d}{dx}\frac{\partial F}{\partial y''} + f'\right)\delta y\right]\bigg|_{x=b},$$

$$\psi(a) = \left[\left(\frac{\partial F}{\partial y'} - \frac{d}{dx}\frac{\partial F}{\partial y''} + g'\right)\delta y\right]\bigg|_{x=a}$$

$$\gamma(b) = \left[\left(\frac{\partial F}{\partial y''} + h'\right)\delta y'\right]\bigg|_{x=b}, \qquad \mu(a) = \left[\left(\frac{\partial F}{\partial y''} + p'\right)\delta y'\right]\bigg|_{x=a}$$

The 1st order necessary condition $\delta J = 0$ provides the necessary equations for a local minimum of J. First choose $\delta y \in C(I)$ such that $\delta y(a) = \delta y(b) = \delta y'(a) = \delta y'(b) = 0$, then

$$\phi(b) = \psi(a) = \gamma(b) = \mu(a) = 0$$

We now require $\delta J_1 = 0$. The Fundamental Lemma 4.6.1 implies the term in brackets in δJ_1 must vanish. Hence, the differential equation (6.1.4) must hold. We now have that $\delta J_1 = 0$ for any δy. The only way the remaining functionals can vanish is if the following boundary conditions are satisfied:

$$\left[\frac{\partial F}{\partial y'} - \frac{d}{dx}\frac{\partial F}{\partial y''} + f'\right]\bigg|_{x=b} = 0, \qquad \left[\frac{\partial F}{\partial y'} - \frac{d}{dx}\frac{\partial F}{\partial y''} + g'\right]\bigg|_{x=a} = 0$$

$$\left[\frac{\partial F}{\partial y''} + h'\right]\bigg|_{x=b} = 0, \qquad \left[\frac{\partial F}{\partial y''} + p'\right]\bigg|_{x=a} = 0$$

We have proved the following theorem.

Theorem 6.2.1. *Consider the closed interval* $I = [a, b]$, *the set of functions* $\mathcal{U} = C^2(I)$, *and the functional* $J : \mathcal{U} \to \mathbb{R}$ *defined by*

$$J[y] = \int_a^b F(x, y, y', y'')\,dx + f(y(b)) - g(y(a)) + h(y'(b)) - p(y'(a)) \qquad (6.2.4)$$

where F has continuous partial derivatives up to order 3 and $f, g, h, p \in C^1(\mathbb{R})$. A necessary condition for a local minimum of J over $\mathcal{U}$ is that the following equations be satisfied:

$$\frac{\partial F}{\partial y} - \frac{d}{dx}\frac{\partial F}{\partial y'} + \frac{d^2}{dx^2}\frac{\partial F}{\partial y''} = 0 \qquad (6.2.5)$$

$$\left[\frac{\partial F}{\partial y'} - \frac{d}{dx}\frac{\partial F}{\partial y''} + f'\right]\bigg|_{x=b} = 0, \qquad \left[\frac{\partial F}{\partial y'} - \frac{d}{dx}\frac{\partial F}{\partial y''} + g'\right]\bigg|_{x=a} = 0 \qquad (6.2.6)$$

$$\left[\frac{\partial F}{\partial y''} + h'\right]\bigg|_{x=b} = 0, \qquad \left[\frac{\partial F}{\partial y''} + p'\right]\bigg|_{x=a} = 0 \qquad (6.2.7)$$

Suppose that $f, g, h, p \in C^2(\mathbb{R})$. Then the 2nd variation is computed according to

$$\delta^2 J = \int_a^b \delta^2 F \, dx + \delta^2[f(y(b))] - \delta^2[g(y(a))] + \delta^2[h(y'(b))] - \delta^2[p(y'(a))]$$

The result is

$$\begin{aligned}
\delta^2 J &= \frac{1}{2}\int_a^b \left[\frac{\partial^2 F}{\partial y^2}(\delta y)^2 + \frac{\partial^2 F}{\partial (y')^2}(\delta y')^2 + \frac{\partial^2 F}{\partial (y'')^2}(\delta y'')^2\right] dx \\
&+ \int_a^b \left[\frac{\partial^2 F}{\partial y \partial y'}\delta y\, \delta y' + \frac{\partial^2 F}{\partial y' \partial y''}\delta y'\, \delta y'' + \frac{\partial^2 F}{\partial y \partial y''}\delta y\, \delta y''\right] dx \\
&+ [f''(y(b)) - g''(y(a))](\delta y)^2 \\
&+ [h''(y'(b)) - p''(y'(a))](\delta y')^2
\end{aligned} \qquad (6.2.8)$$

This page intentionally left blank.

6.3 Cantilever Beam

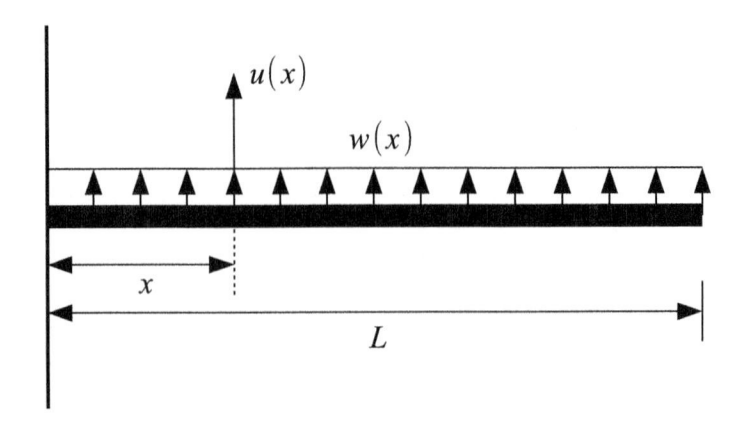

Figure 6.3.1: Cantilever beam

Figure 6.3.1 shows a prismatic cantilever beam of length L with a distributed lateral load $w(x)$. The lateral displacement of the beam is $u(x)$. For small displacements, the angle of deflection is given by $u'(x)$.

The equilibrium displacement $u(x)$ is the one that minimizes the total potential energy Π.

$$\Pi[u] = U(u'') - V(u)$$
$$U(u'') = \frac{1}{2}EI \int_0^L [u''(x)]^2 dx$$
$$V(u) = \int_0^L w(x)\,u(x)\,dx$$

where U is the strain energy and V is the work done by external forces. E is the modulus of elasticity and I is the moment of inertia. We combine the energy terms into a single integral

$$\Pi[u] = \int_0^L \left\{ \frac{1}{2}EI[u''(x)]^2 - w(x)\,u(x) \right\} dx \qquad (6.3.1)$$

The distributed load is assumed to be continuous: $w \in \mathcal{C}([0, L])$. The displacement and angle of displacement are continuous in order to meet physical requirements. We have the boundary conditions $u(0) = 0$ and $u'(0) = 0$. The solution space is formulated as

$$\mathcal{U} = \left\{ u \in \mathcal{C}^2([0, L]) \mid u(0) = 0,\ u'(0) = 0 \right\}$$

No variation is allowed at $x = 0$ and the variational space is: $\mathcal{V} = \mathcal{U}$.

6.3. CANTILEVER BEAM

In order to use the results of the previous section, we formulate the functional Π as

$$\Pi[u] = \int_0^L \tilde{F}(u, u'') \, dx$$

$$\tilde{F}(u, u'') = \frac{1}{2} EI(u'')^2 - wu$$

The necessary differential equation is computed as

$$\frac{\partial \tilde{F}}{\partial u} - \frac{d}{dx}\frac{\partial \tilde{F}}{\partial u'} + \frac{d^2}{dx^2}\frac{\partial \tilde{F}}{\partial u''} = 0$$

$$-w + EI\, u^{(iv)} = 0$$

The natural boundary conditions at $x = L$ are

$$\left.\frac{\partial \tilde{F}}{\partial u''}\right|_L = 0, \qquad \left[\frac{\partial \tilde{F}}{\partial u'} - \frac{d}{dx}\frac{\partial \tilde{F}}{\partial u''}\right]_L = 0$$

Computing the partial derivatives gives

$$EI\, u''(L) = 0, \qquad -EI\, u'''(L) = 0$$

The results above give the differential equation of beam bending

$$EI\, u^{(iv)}(x) = w(x) \tag{6.3.2}$$

with the boundary conditions:

$$u(0) = u'(0) = 0, \qquad u''(L) = u'''(L) = 0 \tag{6.3.3}$$

An alternate derivation of the differential equation is provided in structural mechanics texts. Those derivations use kinematic assumptions and material science. More details can be found in [11, 12, 14, 16, 20].

The 2nd variation is computed as follows:

$$\begin{aligned}
\delta^2 \Pi &= \frac{1}{2}\int_0^L \left[\frac{\partial^2 \tilde{F}}{\partial u^2}(\delta u)^2 + \frac{\partial^2 \tilde{F}}{\partial (u')^2}(\delta u')^2 + \frac{\partial^2 \tilde{F}}{\partial (u'')^2}(\delta u'')^2\right] dx \\
&+ \int_0^L \left[\frac{\partial^2 \tilde{F}}{\partial u \partial u'}\delta u\, \delta u' + \frac{\partial^2 \tilde{F}}{\partial u' \partial u''}\delta u'\, \delta u'' + \frac{\partial^2 \tilde{F}}{\partial u \partial u''}\delta u\, \delta u''\right] dx \\
&= \frac{1}{2}\int_0^L \left[EI\, (\delta u'')^2\right] dx \\
&= \frac{1}{2} EI \int_0^L (\delta u'')^2 dx
\end{aligned}$$

The 2nd variation will be shown to be strictly positive for all admissible non-zero variations. We already have

$$\delta^2 \Pi = \frac{1}{2} EI \int_0^L (\delta u'')^2 dx \geq 0$$

$\delta^2 \Pi = 0$ only when $\delta u''(x) = 0$ for all $x \in [0, L]$ because $\delta u \in \mathcal{V} \subset \mathcal{C}^2([a, b])$. In this case, we integrate twice to get $\delta u(x) = c_1 x + c_0$ for some constants c_0 and c_1. However, $\delta u(0) = \delta u'(x) = 0$ implies $c_1 = c_0 = 0$. Thus for all non-zero variations, $\delta^2 \Pi > 0$ and this confirms a local minimum. The minimum is global because $\delta^2 \Pi > 0$ does not depend on u.

The physical interpretation of $\delta u''(x) = 0$ will now be explained. Remember that $\mathcal{U} = \mathcal{V}$ and we can set $\delta u(x) = u(x)$. If $u(x) = c_0$, then we have rigid body translation. For small c_1, the displacement $u(x) = c_1 x$ implies the beam undergoes a rigid body rotation. Combined translation and rotation is given by the form $u(x) = c_1 x + c_0$. Rigid body motions must be excluded for a meaningful solution to be obtained.

Part III

Calculus of Variations for Multiple Functions

Chapter 7

Formulation

7.1 Introduction

There exist problems in the calculus of variations which involve functionals containing multiple independent functions and their derivatives. The functionals have the form

$$J[y_1,\ldots,y_n] = \int_a^b F(x, y_1, \ldots, y_n, y_1', \ldots, y_n', y_1'', \ldots, y_n'') \, dx \qquad (7.1.1)$$

The function F has continuous partial derivatives up to order 2 in each of its arguments. Each function $y_i(x)$ belongs some appropriate function space on the interval $I = [a, b]$. The form of F in equation (7.1.1) is very cumbersome to write. We introduce the vector-valued function $\mathbf{y} : I \to \mathbb{R}^n$ by defining: $\mathbf{y}(x) = [y_1(x), \ldots, y_n(x)]^T$. Its corresponding derivatives are $\mathbf{y}'(x) = [y_1'(x), \ldots, y_n'(x)]^T$ and $\mathbf{y}''(x) = [y_1''(x), \ldots, y_n''(x)]^T$. Using this notation, equation (7.1.1) is now written as

$$J[\mathbf{y}] = \int_a^b F(x, \mathbf{y}, \mathbf{y}', \mathbf{y}'') \, dx \qquad (7.1.2)$$

We are interested in formulating necessary conditions for a local minimum of the functional (7.1.2). The local minimum problem is formally stated as

$$\mathbf{y}^* = \arg \operatorname*{local\,min}_{\mathbf{y} \in \mathcal{U}} J[\mathbf{y}] \qquad (7.1.3)$$

where $\mathcal{U}$ is some suitable function space. The proofs of the results in the remainder of this book follow the exact procedures of previous chapters. The reader may wish to skip these details and focus on the results.

7.2 General Functionals

Calculus of variations require the formulation of both the solution space $\mathcal{U}$ and space of admissible variations $\mathcal{V}$. Since we are now dealing with vector-valued functions, some further definitions are needed. $\mathcal{C}^{(n,m)}(I)$ is the set of vector-valued functions $\mathbf{y}(x) = [y_1(x), \ldots, y_n(x)]^T$ on I where each $y_i(x) \in \mathcal{C}^m(I)$. The n in $\mathcal{C}^{(n,m)}$ refers to the number of functions and m is the order of continuity of each function. The $(n,m)sup$ norm for the Banach space $\mathcal{C}^{(n,m)}(I)$ is

$$\|\mathbf{y}\|_{(n,m)sup} = \sum_{i=1}^{n} \|y_i\|_{msup} \qquad (7.2.1)$$

A variation $\delta\mathbf{y} = [\delta y_1(x), \ldots, \delta y_n(x)]^T$ is *admissible* if $\mathbf{y} + \delta\mathbf{y} \in \mathcal{U}$. The specification of the sets of possible solutions $\mathcal{U}$ and admissible variations $\mathcal{V}$ depends on the boundary conditions of the problem. However, we require that $\mathcal{U} \subset \mathcal{C}^{(n,m)}(I)$ and $\mathcal{V}$ be a subspace of $\mathcal{C}^{(n,m)}(I)$. We now define the 1st variation of the functional (7.1.2).

Definition 7.2.1 (1st Variation). The functional $J[\mathbf{y}]$ has a *1st variation* if there exists a linear functional $\delta J[\mathbf{y}] : \mathcal{V} \to \mathbb{R}$ such that

$$J[\mathbf{y} + \delta\mathbf{y}] = J[\mathbf{y}] + \delta J[\mathbf{y}][\delta\mathbf{y}] + \varepsilon \|\delta\mathbf{y}\|_{(n,m)sup} \qquad (7.2.2)$$

where $\mathbf{y} + \delta\mathbf{y} \in \mathcal{U}$ and $\varepsilon \to 0$ as $\|\delta\mathbf{y}\|_{(n,m)sup} \to 0$.

The results of the single function case are repeated in this section for the sake of completeness. The wording is identical and the only change is that we now deal with with vector-valued functions from the Banach space $\mathcal{C}^{(n,m)}$.

Scalar multiplication and vector addition for functionals are defined in the usual manner. Let $\alpha \in \mathbb{R}$ and J and K be functionals on $\mathcal{U}$.

$$\begin{aligned}(\alpha J)[\mathbf{y}] &= \alpha J[\mathbf{y}] \\ (J + K)[\mathbf{y}] &= J[\mathbf{y}] + K[\mathbf{y}]\end{aligned}$$

We show that the 1st variation is a linear operator.

7.2. GENERAL FUNCTIONALS

Lemma 7.2.1. *The 1st variation is a linear operator. If J and K are functionals on $\mathcal{U}$ with a 1st variation and $\alpha \in \mathbb{R}$, then for all $\mathbf{y} \in \mathcal{U}$ and admissible variations $\delta\mathbf{y} \in \mathcal{V}$*

$$\delta(\alpha J)[\mathbf{y}][\delta\mathbf{y}] = \alpha\,\delta J[\mathbf{y}][\delta\mathbf{y}]$$
$$\delta(J+K)[\mathbf{y}][\delta\mathbf{y}] = \delta J[\mathbf{y}][\delta\mathbf{y}] + \delta K[\mathbf{y}][\delta\mathbf{y}]$$

Proof. Using the fact that $(\alpha J)[\mathbf{y}] = \alpha\, J[\mathbf{y}]$ and the definition of 1st variation of J, we write

$$\begin{aligned}
(\alpha J)[\mathbf{y}+\delta\mathbf{y}] &= \alpha\, J[\mathbf{y}+\delta\mathbf{y}] \\
&= \alpha\left\{J[\mathbf{y}] + \delta J[\mathbf{y}][\delta\mathbf{y}] + \varepsilon\,\|\delta\mathbf{y}\|_{(n,m)sup}\right\} \\
&= \alpha\, J[\mathbf{y}] + \alpha\,\delta J[\mathbf{y}][\delta\mathbf{y}] + \alpha\varepsilon\,\|\delta\mathbf{y}\|_{(n,m)sup}
\end{aligned}$$

where $\varepsilon \to 0$ as $\|\delta\mathbf{y}\|_{(n,m)sup} \to 0$. Now define $\tilde{\varepsilon} = \alpha\varepsilon$ and write

$$(\alpha J)[\mathbf{y}+\delta\mathbf{y}] = (\alpha J)[\mathbf{y}] + \alpha\,\delta J[\mathbf{y}][\delta\mathbf{y}] + \tilde{\varepsilon}\,\|\delta\mathbf{y}\|_{(n,m)sup}$$

We have $\tilde{\varepsilon} \to 0$ as $\|\delta\mathbf{y}\|_{(n,m)sup} \to 0$. The 1st variation must be defined as $\delta(\alpha J)[\mathbf{y}][\delta\mathbf{y}] = \alpha\,\delta J[\mathbf{y}][\delta\mathbf{y}]$.

The linearity property of addition is proved by using $(J+K)[\mathbf{y}] = J[\mathbf{y}]+K[\mathbf{y}]$.

$$\begin{aligned}
(J+K)[\mathbf{y}+\delta\mathbf{y}] &= J[\mathbf{y}+\delta\mathbf{y}] + K[\mathbf{y}+\delta\mathbf{y}] \\
&= J[\mathbf{y}] + \delta J[\mathbf{y}][\delta\mathbf{y}] + \varepsilon_1\,\|\delta\mathbf{y}\|_{(n,m)sup} \\
&\quad + K[\mathbf{y}] + \delta K[\mathbf{y}][\delta\mathbf{y}] + \varepsilon_2\,\|\delta\mathbf{y}\|_{(n,m)sup} \\
&= J[\mathbf{y}] + K[\mathbf{y}] + \delta J[\mathbf{y}][\delta\mathbf{y}] + \delta K[\mathbf{y}][\delta\mathbf{y}] \\
&\quad + (\varepsilon_1+\varepsilon_2)\,\|\delta\mathbf{y}\|_{nsup}
\end{aligned}$$

where $\varepsilon_1, \varepsilon_2 \to 0$ as $\|\delta\mathbf{y}\|_{(n,m)sup} \to 0$. Let $\tilde{\varepsilon} = \varepsilon_1 + \varepsilon_2$ and we observe that

$$(J+K)[\mathbf{y}+\delta\mathbf{y}] = (J+K)[\mathbf{y}] + \delta J[\mathbf{y}][\delta\mathbf{y}] + \delta K[\mathbf{y}][\delta\mathbf{y}] + \tilde{\varepsilon}\,\|\delta\mathbf{y}\|_{nsup}$$

Hence $\delta(J+K)[\mathbf{y}][\delta\mathbf{y}] = \delta J[\mathbf{y}][\delta\mathbf{y}] + K[\mathbf{y}][\delta\mathbf{y}]$. □

Lemma 7.2.2. *Let $\mathcal{V}$ be a subspace of $\mathcal{C}^{(n,m)}([a,b])$. If $\varphi : \mathcal{V} \to \mathbb{R}$ is a linear functional such that $\varphi[\mathbf{v}]/\|\mathbf{v}\|_{(n,m)sup} \to 0$ as $\|\mathbf{v}\|_{(n,m)sup} \to 0$, then $\varphi[\mathbf{v}] = 0$ for all $\mathbf{v} \in \mathcal{V}$.*

Proof. The proof is by contradiction. Suppose that $\varphi[\mathbf{v}] \neq 0$. Choose $\mathbf{v}_0 \in \mathcal{V}$ such that $\mathbf{v}_0 \neq \mathbf{0}$ and $\varphi[\mathbf{v}_0] = \varphi_0 \neq 0$. Clearly $\|\mathbf{v}_0\|_{(n,m)sup} \neq 0$. Let $\mathbf{v}_k = \mathbf{v}_0/k$. Then $\mathbf{v}_k \to \mathbf{0}$ as $k \to \infty$. Since $\mathbf{v}_k \in \mathcal{V}$, we also have $\|\mathbf{v}_k\|_{(n,m)sup} \to 0$ as $k \to \infty$.

$$\lim_{k\to\infty} \frac{\varphi[\mathbf{v}_k]}{\|\mathbf{v}_k\|_{(n,m)sup}} = \lim_{k\to\infty} \frac{\varphi[\mathbf{v}_0/k]}{\|\mathbf{v}_0/k\|_{(n,m)sup}}$$

$$= \lim_{k\to\infty} \frac{\frac{1}{k}\varphi[\mathbf{v}_0]}{\frac{1}{k}\|\mathbf{v}_0\|_{(n,m)sup}} = \frac{\varphi_0}{\|\mathbf{v}_0\|_{(n,m)sup}} \neq 0$$

This contradicts the hypothesis of the theorem. Therefore, we must have $\varphi[\mathbf{v}] = 0$ for all $\mathbf{v} \in \mathcal{V}$. $\square$

Lemma 7.2.3. *The 1st variation of a functional is unique.*

Proof. Let δJ_1 and δJ_2 be first variations of the functional J. Then by the definition of 1st variation, we write

$$\begin{aligned} J[\mathbf{y} + \delta\mathbf{y}] &= J[\mathbf{y}] + \delta J_1[\mathbf{y}][\delta\mathbf{y}] + \varepsilon_1 \|\delta\mathbf{y}\|_{(n,m)sup} \\ J[\mathbf{y} + \delta\mathbf{y}] &= J[\mathbf{y}] + \delta J_2[\mathbf{y}][\delta\mathbf{y}] + \varepsilon_2 \|\delta\mathbf{y}\|_{(n,m)sup} \end{aligned}$$

where where $\mathbf{y} + \delta\mathbf{y} \in \mathcal{U}$ and $\varepsilon_1, \varepsilon_2 \to 0$ as $\|\delta\mathbf{y}\|_{(n,m)sup} \to 0$. Subtracting the two equations above and rearranging gives

$$\delta J_1[\mathbf{y}][\delta\mathbf{y}] - \delta J_2[\mathbf{y}][\delta\mathbf{y}] = (\varepsilon_2 - \varepsilon_1) \|\delta\mathbf{y}\|_{(n,m)sup}$$

The left hand side is the linear functional $\varphi = (\delta J_1 - \delta J_2)[\mathbf{y}]$.

$$\varphi[\delta\mathbf{y}] = (\varepsilon_2 - \varepsilon_1) \|\delta\mathbf{y}\|_{(n,m)sup}$$

Divide both sides by $\|\delta\mathbf{y}\|_{(n,m)sup}$ and let $\|\delta\mathbf{y}\|_{(n,m)sup} \to 0$.

$$\frac{\varphi[\delta\mathbf{y}]}{\|\delta\mathbf{y}\|_{(n,m)sup}} \to 0$$

Using the previous lemma, this implies that $\varphi[\delta\mathbf{y}] = 0$ and hence $\delta J_1 = \delta J_2$. $\square$

7.2. GENERAL FUNCTIONALS

Definition 7.2.2 (2nd Variation)**.** The functional $J[\mathbf{y}]$ has a *2nd variation* if it has a first variation and there exists a quadratic functional $\delta^2 J[\mathbf{y}] : \mathcal{V} \to \mathbb{R}$ such that

$$J[\mathbf{y} + \delta\mathbf{y}] = J[\mathbf{y}] + \delta J[\mathbf{y}][\delta\mathbf{y}] + \delta^2 J[\mathbf{y}][\delta\mathbf{y}] + \varepsilon \left\| \delta\mathbf{y} \right\|_{(n,m)sup}^{2} \quad (7.2.3)$$

where $\mathbf{y} + \delta\mathbf{y} \in \mathcal{U}$ and $\varepsilon \to 0$ as $\left\| \delta\mathbf{y} \right\|_{(n,m)sup} \to 0$.

Lemma 7.2.4. *The 2nd variation is a linear operator. If J and K are functions on $\mathcal{U}$ with a 2nd variation and $\alpha \in \mathbb{R}$, then for all $\mathbf{y} \in \mathcal{U}$ and admissible variations $\delta\mathbf{y} \in \mathcal{V}$*

$$\begin{aligned}
\delta^2(\alpha J)[\mathbf{y}][\delta\mathbf{y}] &= \alpha\, \delta^2 J[\mathbf{y}][\delta\mathbf{y}] \\
\delta^2(J + K)[\mathbf{y}][\delta\mathbf{y}] &= \delta^2 J[\mathbf{y}][\delta\mathbf{y}] + \delta^2 K[\mathbf{y}][\delta\mathbf{y}]
\end{aligned}$$

Proof. The definition of 2nd variation of J gives

$$\begin{aligned}
(\alpha J)[\mathbf{y} + \delta\mathbf{y}] &= \alpha\, J[\mathbf{y} + \delta\mathbf{y}] \\
&= \alpha \left\{ J[\mathbf{y}] + \delta J[\mathbf{y}][\delta\mathbf{y}] + \delta^2 J[\mathbf{y}][\delta\mathbf{y}] + \varepsilon \left\| \delta\mathbf{y} \right\|_{(n,m)sup}^{2} \right\} \\
&= \alpha\, J[\mathbf{y}] + \alpha\, \delta J[\mathbf{y}][\delta\mathbf{y}] + \alpha\, \delta^2 J[\mathbf{y}][\delta\mathbf{y}] + \alpha\varepsilon \left\| \delta\mathbf{y} \right\|_{(n,m)sup}^{2}
\end{aligned}$$

where $\varepsilon \to 0$ as $\left\| \delta\mathbf{y} \right\|_{(n,m)sup} \to 0$. The 1st variation is a linear operator and we define $\tilde{\varepsilon} = \alpha\varepsilon$.

$$(\alpha J)[\mathbf{y} + \delta\mathbf{y}] = (\alpha J)[\mathbf{y}] + \delta(\alpha J)[\mathbf{y}][\delta\mathbf{y}] + \alpha\, \delta^2 J[\mathbf{y}][\delta\mathbf{y}] + \tilde{\varepsilon} \left\| \delta\mathbf{y} \right\|_{(n,m)sup}^{2}$$

We have $\tilde{\varepsilon} \to 0$ as $\left\| \delta\mathbf{y} \right\|_{(n,m)sup} \to 0$. The 2nd variation must be defined as $\delta^2(\alpha J)[\mathbf{y}][\delta\mathbf{y}] = \alpha\, \delta^2 J[\mathbf{y}][\delta\mathbf{y}]$.

The linearity property of addition is proved as follows:

$$\begin{aligned}
(J + K)[\mathbf{y} + \delta\mathbf{y}] &= J[\mathbf{y} + \delta\mathbf{y}] + K[\mathbf{y} + \delta\mathbf{y}] \\
&= J[\mathbf{y}] + \delta J[\mathbf{y}][\delta\mathbf{y}] + \delta^2 J[\mathbf{y}][\delta\mathbf{y}] + \varepsilon_1 \left\| \delta\mathbf{y} \right\|_{(n,m)sup}^{2} \\
&\quad + K[\mathbf{y}] + \delta K[\mathbf{y}][\delta\mathbf{y}] + \delta^2 K[\mathbf{y}][\delta\mathbf{y}] + \varepsilon_2 \left\| \delta\mathbf{y} \right\|_{(n,m)sup}^{2} \\
&= J[\mathbf{y}] + K[\mathbf{y}] + \delta J[\mathbf{y}] + \delta K[\mathbf{y}] \\
&\quad + \delta^2 J[\mathbf{y}][\delta\mathbf{y}] + \delta^2 K[\mathbf{y}][\delta\mathbf{y}] + (\varepsilon_1 + \varepsilon_2) \left\| \delta\mathbf{y} \right\|_{(n,m)sup}^{2}
\end{aligned}$$

where $\varepsilon_1, \varepsilon_2 \to 0$ as $\left\| \delta\mathbf{y} \right\|_{(n,m)sup} \to 0$. Let $\tilde{\varepsilon} = \varepsilon_1 + \varepsilon_2$ and notice that

$$\begin{aligned}
(J + K)[\mathbf{y} + \delta\mathbf{y}] &= (J + K)[\mathbf{y}] + \delta(J + K)[\mathbf{y}][\delta\mathbf{y}] \\
&\quad + \delta^2 J[\mathbf{y}][\delta\mathbf{y}] + \delta^2 K[\mathbf{y}][\delta\mathbf{y}] + \tilde{\varepsilon} \left\| \delta\mathbf{y} \right\|_{(n,m)sup}
\end{aligned}$$

Hence $\delta^2(J + K)[\mathbf{y}][\delta\mathbf{y}] = \delta^2 J[\mathbf{y}][\delta\mathbf{y}] + \delta^2 K[\mathbf{y}][\delta\mathbf{y}]$. □

Lemma 7.2.5. *Let $\mathcal{V}$ be a subspace of $\mathcal{C}^{(n,m)}([a,b])$. If $\psi : \mathcal{V} \to \mathbb{R}$ is a quadratic functional such that $\psi[\mathbf{v}]/\|\mathbf{v}\|_{(n,m)sup}^2 \to 0$ as $\|\mathbf{v}\|_{(n,m)sup} \to 0$, then $\psi[\mathbf{v}] = 0$ for all $\mathbf{v} \in \mathcal{V}$.*

Proof. The proof is by contradiction. Suppose that $\psi[\mathbf{v}] \neq 0$. Choose $\mathbf{v}_0 \in \mathcal{V}$ such that $\mathbf{v}_0 \neq \mathbf{0}$ and $\psi[\mathbf{v}_0] = \psi_0 \neq 0$. Clearly $\|\mathbf{v}_0\|_{(n,m)sup} \neq 0$. Let $\mathbf{v}_k = \mathbf{v}_0/k$. Then $\mathbf{v}_k \to \mathbf{0}$ as $k \to \infty$. Since $\mathbf{v}_k \in \mathcal{V}$, we also have $\|\mathbf{v}_k\|_{(n,m)sup} \to 0$ as $k \to \infty$.

$$\lim_{k \to \infty} \frac{\psi[\mathbf{v}_k]}{\|\mathbf{v}_k\|_{(n,m)sup}^2} = \lim_{k \to \infty} \frac{\psi[\mathbf{v}_0/k]}{\|\mathbf{v}_0/k\|_{(n,m)sup}^2}$$

$$= \lim_{k \to \infty} \frac{\frac{1}{k^2}\psi[\mathbf{v}_0]}{\frac{1}{k^2}\|\mathbf{v}_0\|_{(n,m)sup}^2} = \frac{\psi_0}{\|\mathbf{v}_0\|_{(n,m)sup}^2} \neq 0$$

This contradicts the hypothesis of the theorem. Therefore, we must have $\psi[\mathbf{v}] = 0$ for all $\mathbf{v} \in \mathcal{V}$. $\square$

Lemma 7.2.6. *The 2nd variation of a functional is unique.*

Proof. Let $\delta^2 J_1$ and $\delta^2 J_2$ be 2nd variations of the functional J. Then by the definition of 2nd variation, we can write

$$J[\mathbf{y} + \delta\mathbf{y}] = J[\mathbf{y}] + \delta J[\mathbf{y}][\delta\mathbf{y}] + \delta^2 J_1[\mathbf{y}][\delta\mathbf{y}] + \varepsilon_1 \|\delta\mathbf{y}\|_{(n,m)sup}^2$$
$$J[\mathbf{y} + \delta\mathbf{y}] = J[\mathbf{y}] + \delta J[\mathbf{y}][\delta\mathbf{y}] + \delta^2 J_2[\mathbf{y}][\delta\mathbf{y}] + \varepsilon_2 \|\delta\mathbf{y}\|_{(n,m)sup}^2$$

where where $\mathbf{y} + \delta\mathbf{y} \in \mathcal{U}$ and $\varepsilon_1, \varepsilon_2 \to 0$ as $\|\delta\mathbf{y}\|_{(n,m)sup} \to 0$. Subtracting the two equations above and rearranging gives

$$\delta^2 J_1[\mathbf{y}][\delta\mathbf{y}] - \delta^2 J_2[\mathbf{y}][\delta\mathbf{y}] = (\varepsilon_2 - \varepsilon_1) \|\delta\mathbf{y}\|_{(n,m)sup}^2$$

The left hand side is the quadratic functional $\psi = (\delta^2 J_1 - \delta^2 J_2)[\mathbf{y}]$.

$$\psi[\delta\mathbf{y}] = (\varepsilon_2 - \varepsilon_1) \|\delta\mathbf{y}\|_{(n,m)sup}^2$$

Divide both sides by $\|\delta\mathbf{y}\|_{nsup}^2$ and let $\|\delta\mathbf{y}\|_{(n,m)sup} \to 0$.

$$\frac{\psi[\delta\mathbf{y}]}{\|\delta\mathbf{y}\|_{(n,m)sup}^2} \to 0$$

Using the previous lemma, this implies that $\psi[\delta\mathbf{y}] = 0$ and hence $\delta^2 J_1 = \delta^2 J_2$. $\square$

7.2. GENERAL FUNCTIONALS

We repeat the definitions of continuity classes of functionals using the same language as for functionals of a single function. Vector notation and the appropriate norms are used for the multi-function case. Let $\mathcal{U} \subset \mathcal{C}^{(n,m)}([a,b])$ and $\mathcal{V} \subset \mathcal{C}^{(n,m)}([a,b])$.

Definition 7.2.3 (Continuous Functional). $J[\mathbf{y}]$ is a *continuous functional* on $\mathcal{U} \subset \mathcal{C}^{(n,m)}([a,b])$ if for every $\varepsilon > 0$, there exists $\delta > 0$ such that $|J[\mathbf{y}] - J[\mathbf{y}_0]| < \varepsilon$ whenever $\|\mathbf{y} - \mathbf{y}_0\|_{(n,m)sup} < \delta$ where $\mathbf{y}, \mathbf{y}_0 \in \mathcal{U}$.

Definition 7.2.4 (Continuous 1st variation). $\delta J[\mathbf{y}][\delta \mathbf{y}]$ is *continuous* on $\mathcal{U} \times \mathcal{V}$ if it is continuous in each of its arguments. For every $\delta \mathbf{y} \in \mathcal{V}$ and $\varepsilon > 0$, there exists $\delta > 0$ such that $|\delta J[\mathbf{y}][\delta \mathbf{y}] - \delta J[\mathbf{y}_0][\delta \mathbf{y}]| < \varepsilon$ whenever $\|\mathbf{y} - \mathbf{y}_0\|_{(n,m)sup} < \delta$ where $\mathbf{y}, \mathbf{y}_0 \in \mathcal{U}$. Furthermore, $\delta J[\mathbf{y}][\delta \mathbf{y}]$ is a continuous linear functional in $\delta \mathbf{y}$ for every $\mathbf{y} \in \mathcal{U}$.

Definition 7.2.5 (Continuous 2nd variation). $\delta^2 J[\mathbf{y}][\delta \mathbf{y}]$ is *continuous* on $\mathcal{U} \times \mathcal{V}$ if it is continuous in each of its arguments. For every $\delta \mathbf{y} \in \mathcal{V}$ and $\varepsilon > 0$, there exists $\delta > 0$ such that $|\delta^2 J[\mathbf{y}][\delta \mathbf{y}] - \delta^2 J[\mathbf{y}_0][\delta \mathbf{y}]| < \varepsilon$ whenever $\|\mathbf{y} - \mathbf{y}_0\|_{(n,m)sup} < \delta$ where $\mathbf{y}, \mathbf{y}_0 \in \mathcal{U}$. Furthermore, $\delta^2 J[\mathbf{y}][\delta \mathbf{y}]$ is a continuous functional in $\delta \mathbf{y}$ for every $\mathbf{y} \in \mathcal{U}$.

We denote the class of continuous functionals on $\mathcal{U}$ as $\mathcal{D}(\mathcal{U})$. The set of continuous functionals with continuous 1st variations are notated as $\mathcal{D}^1(\mathcal{U})$. Similarly, $\mathcal{D}^2(\mathcal{U})$ are continuous functionals with continuous 1st and 2nd variations. It is easy to conclude that $\mathcal{D}^2(\mathcal{U}) \subset \mathcal{D}^1(\mathcal{U}) \subset \mathcal{D}(\mathcal{U})$.

7.3 Alternate Variational Formulation

The alternate variational formulation for multiple functions is provided in this section. Let $\mathcal{U}$ be the solution space and $\mathcal{V}$ be the variational space. Let $\delta \mathbf{y}(x) = \varepsilon \mathbf{v}(x)$, where $\mathbf{v} \in \mathcal{V}$ and ε is a small parameter. We will have $\delta \mathbf{y} \in \mathcal{V}$ for sufficiently small ε. In other words, we require $\mathbf{y} + \delta \mathbf{y} \in \mathcal{U}$ whenever $\mathbf{y} \in \mathcal{U}$ and $\delta \mathbf{y} \in \mathcal{V}$. The 1st and 2nd order Taylor series expansions of a functional are given by the following lemmas.

Lemma 7.3.1. *If $J \in \mathcal{D}^1(\mathcal{U})$, then for every $\mathbf{y} \in \mathcal{U}$ and every $\delta \mathbf{y} = \varepsilon \mathbf{v} \in \mathcal{V}$, we can write*

$$J[\mathbf{y} + \varepsilon \mathbf{v}] = J[\mathbf{y}] + \varepsilon\, J_1[\mathbf{y}][\mathbf{v}] + \mu \varepsilon \left\| \mathbf{v} \right\|_{(n,m)sup} \qquad (7.3.1)$$

where $\mu \to 0$ as $\varepsilon \to 0$ and J_1 is a linear functional in $\mathbf{v}$.

Proof. The definition of 1st variation in the form of equation (7.2.2) is

$$\begin{aligned} J[\mathbf{y} + \varepsilon \mathbf{v}] &= J[\mathbf{y}] + \delta J[\mathbf{y}][\varepsilon \mathbf{v}] + \mu_1 \left\| \varepsilon \mathbf{v} \right\|_{(n,m)sup} \\ &= J[\mathbf{y}] + \delta J[\mathbf{y}][\varepsilon \mathbf{v}] + \mu_1 \left| \varepsilon \right| \left\| \mathbf{v} \right\|_{(n,m)sup} \end{aligned}$$

$\mu_1 \to 0$ as $\varepsilon \to 0$ because $\left\| \varepsilon \mathbf{v} \right\|_{(n,m)sup} \to 0$ when $\varepsilon \to 0$. We can write

$$J[\mathbf{y} + \varepsilon \mathbf{v}] = J[\mathbf{y}] + \delta J[\mathbf{y}][\varepsilon \mathbf{v}] + \mu \varepsilon \left\| \mathbf{v} \right\|_{(n,m)sup}$$

by defining $\mu = \operatorname{sign}(\varepsilon)\, \mu_1$. Note that $\mu \to 0$ as $\varepsilon \to 0$. The factor of ε can be pulled out of the 1st variation because of the linearity property.

$$J[\mathbf{y} + \varepsilon \mathbf{v}] = J[\mathbf{y}] + \varepsilon\, \delta J[\mathbf{y}][\mathbf{v}] + \mu \varepsilon \left\| \mathbf{v} \right\|_{(n,m)sup}$$

Setting $J_1[\mathbf{y}][\mathbf{v}] = \delta J[\mathbf{y}][\mathbf{v}]$ completes the proof. $\square$

Lemma 7.3.2. *If $J \in \mathcal{D}^2(\mathcal{U})$, then for every $\mathbf{y} \in \mathcal{U}$ and every $\delta \mathbf{y} = \varepsilon \mathbf{v} \in \mathcal{V}$, we can write*

$$J[\mathbf{y} + \varepsilon \mathbf{v}] = J[\mathbf{y}] + \varepsilon\, J_1[\mathbf{y}][\mathbf{v}] + \frac{1}{2}\varepsilon^2\, J_2[\mathbf{y}][\mathbf{v}] + \mu \varepsilon^2 \left\| \mathbf{v} \right\|_{(n,m)sup}^2 \qquad (7.3.2)$$

where $\mu \to 0$ as $\varepsilon \to 0$. J_1 is a linear functional in $\mathbf{v}$ and J_2 is a quadratic functional in $\mathbf{v}$.

7.3. ALTERNATE VARIATIONAL FORMULATION

Proof. The definition of 2nd variation in the form of equation (7.2.3) is

$$\begin{aligned} J[\mathbf{y} + \varepsilon \mathbf{v}] &= J[\mathbf{y}] + \delta J[\mathbf{y}][\varepsilon \mathbf{v}] + \delta^2 J[\mathbf{y}][\varepsilon \mathbf{v}] + \mu \left\| \varepsilon \mathbf{v} \right\|^2_{(n,m)sup} \\ &= J[\mathbf{y}] + \varepsilon \, \delta J[\mathbf{y}][\mathbf{v}] + \varepsilon^2 \, \delta^2 J[\mathbf{y}][\mathbf{v}] + \mu \varepsilon^2 \left\| \mathbf{v} \right\|^2_{(n,m)sup} \end{aligned}$$

$\mu \to 0$ as $\varepsilon \to 0$ because $\left\| \varepsilon \mathbf{v} \right\|_{(n,m)sup} \to 0$ when $\varepsilon \to 0$. Factors of ε are pulled out of the variations according to the linear and quadratic functional properties. The proof is complete by making the following associations:

$$J_1[\mathbf{y}][\mathbf{v}] = \delta J[\mathbf{y}][\mathbf{v}], \qquad J_2[\mathbf{y}][\mathbf{v}] = 2 \, \delta^2 J[\mathbf{y}][\mathbf{v}]$$

□

The alternate variational formulation uses derivatives to extract the 1st and 2nd variations of a functional.

Lemma 7.3.3. *If* $J \in \mathcal{D}^1(\mathcal{U})$, *then*

$$\left. \frac{d}{d\varepsilon} J[\mathbf{y} + \varepsilon \mathbf{v}] \right|_{\varepsilon=0} = J_1[\mathbf{y}][\mathbf{v}] = \delta J[\mathbf{y}][\mathbf{v}], \qquad \forall \mathbf{y} \in \mathcal{U}, \forall \mathbf{v} \in \mathcal{V} \qquad (7.3.3)$$

Proof. Compute the derivative of equation (7.3.1) with respect to ε.

$$\frac{d}{d\varepsilon} J[\mathbf{y} + \varepsilon \mathbf{v}] = J_1[\mathbf{y}][\mathbf{v}] + \mu \left\| \mathbf{v} \right\|_{(n,m)sup}$$

Recall that $\mu \to 0$ as $\varepsilon \to 0$. Hence, the error term vanishes when setting $\varepsilon = 0$.

$$\left. \frac{d}{d\varepsilon} J[\mathbf{y} + \varepsilon \mathbf{v}] \right|_{\varepsilon=0} = J_1[\mathbf{y}][\mathbf{v}] = \delta J[\mathbf{y}][\mathbf{v}]$$

□

Lemma 7.3.4. *If* $J \in \mathcal{D}^2(\mathcal{U})$, *then*

$$\left. \frac{d^2}{d\varepsilon^2} J[\mathbf{y} + \varepsilon \mathbf{v}] \right|_{\varepsilon=0} = J_2[\mathbf{y}][\mathbf{v}] = 2 \, \delta^2 J[\mathbf{y}][\mathbf{v}], \qquad \forall \mathbf{y} \in \mathcal{U}, \forall \mathbf{v} \in \mathcal{V} \qquad (7.3.4)$$

Proof. Compute the 2nd derivative of equation (7.3.2) with respect to ε.

$$\frac{d^2}{d\varepsilon^2} J[\mathbf{y} + \varepsilon \mathbf{v}] = J_2[\mathbf{y}][\mathbf{v}] + 2\mu \left\| \mathbf{v} \right\|^2_{(n,m)sup}$$

Recall that $\mu \to 0$ as $\varepsilon \to 0$. Hence, the error term vanishes when setting $\varepsilon = 0$.

$$\left. \frac{d^2}{d\varepsilon^2} J[\mathbf{y} + \varepsilon \mathbf{v}] \right|_{\varepsilon=0} = J_2[\mathbf{y}][\mathbf{v}] = 2 \, \delta^2 J[\mathbf{y}][\mathbf{v}]$$

□

7.4 Integral Functionals

In this section we compute the 1st and 2nd variation of functionals of the form

$$J[\mathbf{y}] = \int_a^b F(x, \mathbf{y}, \mathbf{y}', \mathbf{y}'') \, dx \qquad (7.4.1)$$

Doing so requires computing expansions of F in terms of variations. We first define the augmented vector $\mathbf{u}$ through

$$\mathbf{u} = [x, y_1, \ldots, y_n, y_1', \ldots, y_n', y_1'', \ldots, y_n'']^T = [x, \mathbf{y}^T, (\mathbf{y}')^T, (\mathbf{y}'')^T]^T$$

and write $F(\mathbf{u}) \equiv F(x, \mathbf{y}, \mathbf{y}', \mathbf{y}'')$. The 1st order expansion of F is provided by equation (3.2.12).

$$F(\mathbf{u} + \delta\mathbf{u}) = F(\mathbf{u}) + \nabla F(\mathbf{u}) \cdot \delta\mathbf{u} + r(\delta\mathbf{u})$$

where $\delta\mathbf{u}$ is

$$\delta\mathbf{u} = [\delta x, \delta y_1, \ldots, \delta y_n, \delta y_1', \ldots, \delta y_n', \delta y_1'', \ldots, \delta y_n'']^T = [\delta x, \delta\mathbf{y}^T, (\delta\mathbf{y}')^T, (\delta\mathbf{y}'')^T]$$

and $r(\delta\mathbf{u})$ is $o(\|\delta\mathbf{u}\|_2)$ as $\delta\mathbf{u} \to \mathbf{0}$. Using variational notation and recalling that $\delta x = 0$,

$$\delta F \equiv \nabla F(\mathbf{u}) \cdot \delta\mathbf{u} = \sum_{i=1}^n \left[\frac{\partial}{\partial y_i} F(\mathbf{u}) \, \delta y_i + \frac{\partial}{\partial y_i'} F(\mathbf{u}) \, \delta y_i' + \frac{\partial}{\partial y_i''} F(\mathbf{u}) \, \delta y_i'' \right]$$

$$F(\mathbf{u} + \delta\mathbf{u}) = F(\mathbf{u}) + \delta F + r(\delta\mathbf{u})$$

The increment in J is computed as

$$\begin{aligned} J[\mathbf{y} + \delta\mathbf{y}] &= \int_a^b F(\mathbf{u} + \delta\mathbf{u}) \, dx \\ &= \int_a^b [F(\mathbf{u}) + \delta F + r(\delta\mathbf{u})] \, dx \\ &= \int_a^b F(\mathbf{u}) \, dx + \int_a^b \delta F \, dx + \int_a^b r(\delta\mathbf{u}) \, dx \\ &= J[\mathbf{y}] + \int_a^b \delta F \, dx + \int_a^b r(\delta\mathbf{u}) \, dx \end{aligned}$$

We claim that the 1st variation of J is given by

$$\delta J \equiv \delta J[\mathbf{y}][\delta\mathbf{y}] = \int_a^b \delta F \, dx \qquad (7.4.2)$$

7.4. INTEGRAL FUNCTIONALS

δJ is linear because the integral of the linear functional δF is also linear. We must show the error term has the correct form and order. We first note the inequality

$$\left| \int_a^b r(\delta \mathbf{u}) \, dx \right| \leq \int_a^b |r(\delta \mathbf{u})| \, dx < \int_a^b \varepsilon(x) \, \|\delta \mathbf{u}\|_2 \, dx$$

$\varepsilon(x) \to 0$ as $\delta \mathbf{u} \to 0$ for every $x \in I = [a,b]$ because $r(\delta \mathbf{u})$ is $o(\|\delta \mathbf{u}\|_2)$. Note that $\varepsilon(x) > 0$. We require an expression involving the $(n,2)sup$ norm of $\delta \mathbf{u}$. We first maximize the squares of the components of $\delta \mathbf{u}$.

$$\int_a^b \varepsilon(x) \, \|\delta \mathbf{u}\|_2 \, dx \;=\; \int_a^b \varepsilon(x) \sqrt{\sum_{i=1}^n (\delta y_i)^2 + \sum_{i=1}^n (\delta y_i')^2 + \sum_{i=1}^n (\delta y_i'')^2} \, dx$$

$$\leq \int_a^b \varepsilon(x) \left[\sum_{i=1}^n \max_{x \in I}(\delta y_i)^2 + \sum_{i=1}^n \max_{x \in I}(\delta y_i')^2 + \sum_{i=1}^n \max_{x \in I}(\delta y_i'')^2 \right]^{1/2} dx$$

Next, we use the triangle inequality.

$$\int_a^b \varepsilon(x) \, \|\delta \mathbf{u}\|_2 \, dx \;\leq\; \int_a^b \varepsilon(x) \left[\sum_{i=1}^n \sqrt{\max_{x \in I}(\delta y_i)^2} + \sum_{i=1}^n \sqrt{\max_{x \in I}(\delta y_i')^2} + \sum_{i=1}^n \sqrt{\max_{x \in I}(\delta y_i'')^2} \right] dx$$

$$\leq \int_a^b \varepsilon(x) \left[\sum_{i=1}^n \max_{x \in I}|\delta y_i| + \sum_{i=1}^n \max_{x \in I}|\delta y_i'| + \sum_{i=1}^n \max_{x \in I}|\delta y_i''| \right] dx$$

$$\leq \int_a^b \varepsilon(x) \sum_{i=1}^n \left[\max_{x \in I}|\delta y_i| + \max_{x \in I}|\delta y_i'| + \max_{x \in I}|\delta y_i''| \right] dx$$

$$\leq \int_a^b \varepsilon(x) \sum_{i=1}^n \|\delta y_i\|_{2sup} \, dx$$

$$\leq \int_a^b \varepsilon(x) \, \|\delta \mathbf{y}\|_{(n,2)sup} \, dx$$

$$\leq \|\delta \mathbf{y}\|_{(n,2)sup} \int_a^b \varepsilon(x) \, dx$$

The right hand sides above are all equal.

Now maximize $\varepsilon(x)$ to get

$$\int_a^b \varepsilon(x) \,\|\delta\mathbf{u}\|_2 \, dx \le \|\delta\mathbf{y}\|_{(n,2)sup} \int_a^b \|\varepsilon\|_{\sup} \, dx = (b-a)\,\|\varepsilon\|_{\sup}\,\|\delta\mathbf{y}\|_{(n,2)sup}$$

Let $\tilde{\varepsilon} = (b-a)\,\|\varepsilon\|_{\sup}$ and we get the desired inequality

$$\left|\int_a^b r(\delta\mathbf{u}) \, dx\right| < \tilde{\varepsilon}\,\|\delta\mathbf{y}\|_{(n,2)sup}$$

$\tilde{\varepsilon} \to 0$ as $\|\delta\mathbf{y}\|_{(n,2)nsup} \to 0$ because $\varepsilon(x) \to 0$ as $\delta\mathbf{u}(x) \to \mathbf{0}$ for every $x \in I$. The error term is the correct form and order for equation (7.4.2) to be the 1st variation of J.

If J contained the 1st derivative but not the 2nd derivative, then we would write $\mathbf{u} = [x, (\mathbf{y})^T, (\mathbf{y}')^T, \mathbf{0}^T]^T$, $\delta\mathbf{u} = [\delta x, \delta(\mathbf{y})^T, \delta(\mathbf{y}')^T, \mathbf{0}^T]^T$, and we replace $\|\delta\mathbf{y}\|_{(n,2)sup}$ with $\|\delta\mathbf{y}\|_{(n,1)sup}$.

We also have the commutative property of variation with integration.

$$\delta J = \delta(J) = \delta \int_a^b F \, dx = \int_a^b \delta F \, dx \qquad (7.4.3)$$

Equation (3.4.1) provides the 2nd order expansion of F.

$$F(\mathbf{u}+\delta\mathbf{u}) = F(\mathbf{u}) + \nabla F(\mathbf{u}) \cdot \delta\mathbf{u} + \frac{1}{2}(\delta\mathbf{u})^T \mathbf{H}(\mathbf{u})\,\delta\mathbf{u} + r(\delta\mathbf{u})$$

where $r(\delta\mathbf{u})$ is $o(\|\delta\mathbf{u}\|_2^2)$ as $\delta\mathbf{u} \to \mathbf{0}$. The 2nd variation of F is

$$\delta^2 F \equiv \frac{1}{2}(\delta\mathbf{u})^T \mathbf{H}(\mathbf{u})\,\delta\mathbf{u}$$

with the Hessian matrix $\mathbf{H}(\mathbf{u})$ defined by

$$\mathbf{H}_{ij}(\mathbf{u}) = \frac{\partial^2}{\partial u_i \partial u_j} F(\mathbf{u})$$

where u_i is the corresponding element of $\mathbf{u}$. For example: $u_1 = x$, $u_2 = y_1$, $u_{n+1} = y_n$. Using this information we write

$$F(\mathbf{u}+\delta\mathbf{u}) = F(\mathbf{u}) + \delta F + \delta^2 F + r(\delta\mathbf{u})$$

7.4. INTEGRAL FUNCTIONALS

The increment in J is computed as

$$\begin{aligned}
J[\mathbf{y} + \delta\mathbf{y}] &= \int_a^b F(\mathbf{u} + \delta\mathbf{u})\,dx \\
&= \int_a^b \left[F(\mathbf{u}) + \delta F + \delta^2 F + r(\delta\mathbf{u})\right] dx \\
&= \int_a^b F(\mathbf{u})\,dx + \int_a^b \delta F\,dx + \int_a^b \delta^2 F\,dx + \int_a^b r(\delta\mathbf{u})\,dx \\
&= J[\mathbf{y}] + \delta J + \int_a^b \delta^2 F\,dx + \int_a^b r(\delta\mathbf{u})\,dx
\end{aligned}$$

The 2nd variation of J is

$$\delta^2 J \equiv \delta^2 J[\mathbf{y}][\delta \mathbf{y}] = \int_a^b \delta^2 F\,dx \qquad (7.4.4)$$

$\delta^2 J$ is a quadratic functional because δF is a quadratic functional of $\delta \mathbf{u}$.

The error term for the 2nd variation must be analyzed. We again use the inequality

$$\left|\int_a^b r(\delta \mathbf{u})\,dx\right| \le \int_a^b |r(\delta \mathbf{u})|\,dx < \int_a^b \varepsilon(x)\,\|\delta \mathbf{u}\|_2^2\,dx$$

$\varepsilon(x) \to 0$ as $\delta \mathbf{u} \to \mathbf{0}$ for every $x \in I = [a,b]$ because $r(\delta \mathbf{u})$ is $o(\|\delta \mathbf{u}\|_2^2)$.

Transformation to the $(n,2)sup$ norm of $\delta \mathbf{u}$ starts with maximizing the squares of the components of $\delta \mathbf{u}$.

$$\int_a^b \varepsilon(x)\,\|\delta \mathbf{u}\|_2^2\,dx \le \int_a^b \varepsilon(x) \left[\sum_{i=1}^n \max_{x \in I}(\delta y_i)^2 + \sum_{i=1}^n \max_{x \in I}(\delta y_i')^2 + \sum_{i=1}^n \max_{x \in I}(\delta y_i'')^2\right] dx$$

The right hand side is bounded by the next expression below.

$$\begin{aligned}
\int_a^b \varepsilon(x)\,\|\delta \mathbf{u}\|_2^2\,dx &\le \int_a^b \varepsilon(x) \left[\sum_{i=1}^n \max_{x \in I} |\delta y_i| + \sum_{i=1}^n \max_{x \in I} |\delta y_i'| + \sum_{i=1}^n \max_{x \in I} |\delta y_i''|\right]^2 dx \\
&\le \int_a^b \varepsilon(x) \left\{\sum_{i=1}^n \left[\max_{x \in I} |\delta y_i| + \max_{x \in I} |\delta y_i'| + \max_{x \in I} |\delta y_i''|\right]\right\}^2 dx \\
&\le \int_a^b \varepsilon(x) \left[\sum_{i=1}^n \|\delta y_i\|_{2sup}\right]^2 dx \\
&\le \int_a^b \varepsilon(x)\,\|\delta \mathbf{y}\|_{(n,2)sup}^2\,dx \\
&\le \|\delta \mathbf{y}\|_{(n,2)sup}^2 \int_a^b \varepsilon(x)\,dx
\end{aligned}$$

The last five right hand sides are equal. Maximizing $\varepsilon(x)$ gives

$$\int_a^b \varepsilon(x) \, \|\delta \mathbf{u}\|_2^2 \, dx \leq \|\delta \mathbf{y}\|_{(n,2)sup}^2 \int_a^b \|\varepsilon\|_{\sup} \, dx = (b-a) \, \|\varepsilon\|_{\sup} \, \|\delta \mathbf{y}\|_{(n,2)sup}^2$$

Set $\tilde{\varepsilon} = (b-a) \, \|\varepsilon\|_{\sup}$ to obtain

$$\left| \int_a^b r(\delta \mathbf{u}) \, dx \right| < \tilde{\varepsilon} \, \|\delta \mathbf{y}\|_{(n,2)sup}^2$$

$\tilde{\varepsilon} \to 0$ as $\|\delta \mathbf{y}\|_{(n,2)sup} \to 0$ because $\varepsilon(x) \to 0$ as $\delta \mathbf{u}(x) \to \mathbf{0}$ for every $x \in I$. This inequality provides the correct form and order for the error term in equation (7.4.4) to be the 2nd variation of J.

If the functional J contained the 1st derivative but not the 2nd derivative. we use $\mathbf{u} = [x, \, (\mathbf{y})^T, \, (\mathbf{y}')^T, \, \mathbf{0}^T]^T$, $\delta \mathbf{u} = [\delta x, \, (\delta \mathbf{y})^T, \, (\delta \mathbf{y}')^T, \, 0]^T$, and $\|\delta \mathbf{y}\|_{(n,1)sup}$.

The 2nd variation operator commutes with integration.

$$\delta^2 J = \delta^2(J) = \delta^2 \int_a^b F \, dx = \int_a^b \delta^2 F \, dx \qquad (7.4.5)$$

7.5 Local Minimums

Consider the closed interval $I = [a, b]$. The concept of an open ball in the set $\mathcal{U} \subset \mathcal{C}^{(n,m)}(I)$ is provided by the following definition.

Definition 7.5.1 (Open Ball). The *open ball* $B(\mathbf{y}_0, \delta)$ centered at $\mathbf{y}_0$ with radius δ is the set of all vector-valued functions $\mathbf{y} \in \mathcal{U}$ such that $\|\mathbf{y} - \mathbf{y}_0\|_{(n,m)sup} < \delta$.

The functions $\mathbf{y}^*$ which cause the 1st variation δJ to vanish are called stationary points.

Definition 7.5.2 (Stationary Point). $\mathbf{y}^*$ is the location of a *stationary point* of the functional $J[\mathbf{y}]$ if $\delta J[\mathbf{y}^*][\delta \mathbf{y}] = 0$ for all admissible variations $\delta \mathbf{y}$.

Being a stationary point is a necessary condition for a local minimum.

Definition 7.5.3 (Local Minimum). $\mathbf{y}^*$ is the location of a *local minimum* of $J[\mathbf{y}]$ if there exists $\delta > 0$ such that $J[\mathbf{y}^*] \leq J[\mathbf{y}]$ for all $\mathbf{y} \in B(\mathbf{y}^*, \delta)$.

The necessary conditions for local minimums are provided below.

Theorem 7.5.1 (1st Order Necessary Condition). *If $\mathbf{y}^*$ is the location of a local minimum of $J[\mathbf{y}]$ and $J \in \mathcal{D}^1(\mathcal{U})$, then $\delta J[\mathbf{y}^*][\delta \mathbf{y}] = 0$ for all admissible variations $\delta \mathbf{y}$.*

Proof. The proof is by contradiction. Suppose that $\delta J[\mathbf{y}^*][\delta \mathbf{y}] \neq 0$ for some admissible variation $\delta \mathbf{y} \neq \mathbf{0}$. The definition of 1st variation states that

$$J[\mathbf{y}^* + \alpha\, \delta \mathbf{y}] = J[\mathbf{y}^*] + \delta J[\mathbf{y}^*][\alpha\, \delta \mathbf{y}] + \varepsilon \left\| \alpha\, \delta \mathbf{y} \right\|_{(n,m)sup}$$

for all $\alpha \in [0, 1]$. The error term approaches 0 faster than the 1st variation. If this was not true, then lemma 7.2.2 implies the 1st variation would vanish. Thus, there exists $\delta > 0$ such that $|\varepsilon| \|\alpha\, \delta \mathbf{y}\|_{(n,m)sup} < \frac{1}{2} |\delta J[\mathbf{y}^*][\alpha\, \delta \mathbf{y}]|$ for all $\|\alpha\, \delta \mathbf{y}\|_{(n,m)sup} < \delta$. Set $\bar{\alpha} = \delta / \|\delta \mathbf{y}\|_{(n,m)sup}$ as the upper bound for the size of α. If the 1st variation is negative, then

$$J[\mathbf{y}^* + \alpha\, \delta \mathbf{y}] \leq J[\mathbf{y}^*] + \frac{1}{2} \delta J[\mathbf{y}^*][\alpha\, \delta \mathbf{y}] = J[\mathbf{y}^*] + \frac{1}{2} \alpha\, \delta J[\mathbf{y}^*][\delta \mathbf{y}] < J[\mathbf{y}^*]$$

for all $\alpha \in (0, \bar{\alpha})$. If the 1st variation is positive, then choose $\hat{\alpha} \in (0, \bar{\alpha}]$ small enough so that $\mathbf{y}^* - \alpha\, \delta \mathbf{y} \in \mathcal{U}$. We now have

$$J[\mathbf{y}^* - \alpha\, \delta \mathbf{y}] \leq J[\mathbf{y}^*] + \frac{1}{2} \delta J[\mathbf{y}^*][-\alpha\, \delta \mathbf{y}] = J[\mathbf{y}^*] - \frac{1}{2} \alpha\, \delta J[\mathbf{y}^*][\delta \mathbf{y}] < J[\mathbf{y}^*]$$

for all $\alpha \in (0, \hat{\alpha})$. In either case we contradict that $\mathbf{y}^*$ is a local minimum. Thus the 1st variation must be 0. □

Theorem 7.5.2 (2nd Order Necessary Condition). *If $\mathbf{y}^*$ is the location of a local minimum of $J[\mathbf{y}]$ and $J \in \mathcal{D}^2(\mathcal{U})$, then $\delta J[\mathbf{y}^*][\delta\mathbf{y}] = 0$ and $\delta^2 J[\mathbf{y}^*][\delta\mathbf{y}] \geq 0$ for all admissible variations $\delta\mathbf{y}$.*

Proof. The proof is by contradiction. Suppose that $\delta^2 J[\mathbf{y}^*][\delta\mathbf{y}] < 0$ for some admissible variation $\delta\mathbf{y} \neq \mathbf{0}$. The 1st variation vanishes due to the 1st order necessary condition. The definition of 2nd variation at y^* gives

$$J[\mathbf{y}^* + \alpha\, \delta\mathbf{y}] = J[\mathbf{y}^*] + \delta^2 J[\mathbf{y}^*][\alpha\, \delta\mathbf{y}] + \varepsilon \, \|\alpha\, \delta\mathbf{y}\|^2_{(n,m)sup}$$

for all $\alpha \in [0,1]$. The error term approaches 0 faster than the 2nd variation. Otherwise, lemma 7.2.5 states that the 2nd variation vanishes. Thus, there exists $\delta > 0$ such that $|\varepsilon| \, \|\alpha\, \delta\mathbf{y}\|^2_{(n,m)sup} < \frac{1}{2} \left|\delta^2 J[\mathbf{y}^*][\alpha\, \delta\mathbf{y}]\right|$ for all $\|\alpha\, \delta\mathbf{y}\|_{(n,m)sup} < \delta$. Set $\bar{\alpha} = \delta/\|\delta\mathbf{y}\|_{(n,m)sup}$ as the upper bound for the size of α.

$$J[\mathbf{y}^* + \alpha\, \delta\mathbf{y}] \leq J[\mathbf{y}^*] + \frac{1}{2}\delta^2 J[\mathbf{y}^*][\alpha\, \delta\mathbf{y}] = J[\mathbf{y}^*] + \frac{1}{2}\alpha^2\, \delta^2 J[\mathbf{y}^*][\delta\mathbf{y}] < J[\mathbf{y}^*]$$

for all $\alpha \in (0, \bar{\alpha})$. This contradicts that $\mathbf{y}^*$ is a local minimum. Thus the 2nd variation must be non-negative. $\square$

Theorem 7.5.3 (2nd Order Sufficient Condition). *If $J \in \mathcal{D}^2(\mathcal{U})$, $\delta J[\mathbf{y}^*][\delta\mathbf{y}] = 0$, and $\delta^2 J[\mathbf{y}^*][\delta\mathbf{y}] > 0$ for all admissible variations $\delta\mathbf{y} \neq \mathbf{0}$, then $\mathbf{y}^*$ is the location of a local minimum of $J[\mathbf{y}]$.*

Proof. By the same reasoning in the proof of the 2nd order necessary condition, there exists $\delta > 0$ such that $|\varepsilon| \, \|\delta\mathbf{y}\|^2_{(n,m)sup} < \frac{1}{2}\delta^2 J[\mathbf{y}^*][\delta\mathbf{y}]$ for all $\|\delta\mathbf{y}\|_{(n,m)sup} < \delta$.

$$J[\mathbf{y}^* + \delta\mathbf{y}] = J[\mathbf{y}^*] + \delta^2 J[\mathbf{y}^*][\delta\mathbf{y}] + \varepsilon \, \|\delta\mathbf{y}\|^2_{(n,m)sup} > J[\mathbf{y}^*] + \frac{1}{2}\delta^2 J[\mathbf{y}^*][\delta\mathbf{y}] > J[\mathbf{y}^*]$$

The strict inequality is due to the fact that $\delta\mathbf{y} \neq \mathbf{0}$. Thus $\mathbf{y}^*$ is the location of a local minimum. $\square$

Chapter 8

Functions with 1st Derivatives

8.1 Fixed End Points

In this section, we explore the necessary conditions for minimizing integral functionals of the form

$$J[y_1,\ldots,y_n] = \int_a^b F(x,y_1,\ldots,y_n,y_1',\ldots,y_n')\,dx \qquad (8.1.1)$$

where F has continuous partial derivatives up to order 2 in each of its arguments. We prefer the vector notation

$$J[\mathbf{y}] = \int_a^b F(x,\mathbf{y},\mathbf{y}')\,dx \qquad (8.1.2)$$

where $\mathbf{y}(x) = [y_1(x),\ldots,y_n(x)]^T$ and $\mathbf{y}'(x) = [y_1'(x),\ldots,y_n'(x)]^T$ are vector-valued functions. Let $I = [a,b]$. The set of possible solutions $\mathcal{U}$ is defined as

$$\mathcal{U} = \left\{\mathbf{y} : I \to \mathbb{R}^n \mid \mathbf{y}(a) = \mathbf{y}_a,\ \mathbf{y}(b) = \mathbf{y}_b,\ y_i \in \mathcal{C}^1(I),\ i=1,\ldots,n\right\}$$

The function values at the endpoints are known constants. The local minimization problem is

$$\mathbf{y}^* = \arg\operatorname*{local\,min}_{\mathbf{y}\in\mathcal{U}} J[\mathbf{y}] \qquad (8.1.3)$$

Problem (8.1.3) is solved using the variational procedures developed in previous chapters. We shall see that each function y_i will satisfy the Euler differential equation (5.1.5).

No variation is allowed at the endpoints because the possible solution functions have definite values at those locations. Thus we define

$$\mathcal{V} = \left\{ \delta\mathbf{y} : I \to \mathbb{R}^n \,\middle|\, \delta y_i \in \mathcal{C}_0^1(I), \quad i = 1, \ldots, n \right\}$$

Let $F \equiv F(x, \mathbf{y}, \mathbf{y}')$. We compute the 1st variation using the commutative property (7.4.3) of variation with integration.

$$\delta J = \delta \int_a^b F \, dx = \int_a^b \delta F \, dx$$

The variation of F was defined in the previous chapter and we have

$$\delta J = \int_a^b \sum_{i=1}^n \left[\frac{\partial F}{\partial y_i} \delta y_i + \frac{\partial F}{\partial y_i'} \delta y_i' \right] dx$$

We can always interchange a finite sum with an integral.

$$\delta J = \sum_{i=1}^n \int_a^b \left[\frac{\partial F}{\partial y_i} \delta y_i + \frac{\partial F}{\partial y_i'} \delta y_i' \right] dx \tag{8.1.4}$$

We are now ready to prove the main result for multiple functions with fixed endpoints.

Theorem 8.1.1 (Euler Differential Equations). *Consider the closed interval $I = [a, b]$, the set of functions*

$$\mathcal{U} = \left\{ \mathbf{y} : I \to \mathbb{R}^n \,\middle|\, \mathbf{y}(a) = \mathbf{y}_a, \, \mathbf{y}(b) = \mathbf{y}_b, \quad y_i \in \mathcal{C}^1(I), \quad i = 1, \ldots, n \right\}$$

and the functional $J : \mathcal{U} \to \mathbb{R}$ defined by

$$J[\mathbf{y}] = \int_a^b F(x, \mathbf{y}, \mathbf{y}') \, dx \tag{8.1.5}$$

where F has continuous partial derivatives up to order 2. A necessary condition for a local minimum of J over $\mathcal{U}$ is that the Euler differential equations be satisfied.

$$\frac{\partial F}{\partial y_i} - \frac{d}{dx} \frac{\partial F}{\partial y_i'} = 0, \quad i = 1, \ldots, n \tag{8.1.6}$$

8.1. FIXED END POINTS

Proof. The second term in brackets in equation (8.1.4) is integrated by parts.

$$\begin{aligned}
\delta J &= \sum_{i=1}^{n} \int_{a}^{b} \left[\frac{\partial F}{\partial y_i} \delta y_i + \frac{\partial F}{\partial y_i'} \delta y_i' \right] dx \\
&= \sum_{i=1}^{n} \left\{ \int_{a}^{b} \frac{\partial F}{\partial y_i} \delta y_i \, dx + \int_{a}^{b} \frac{\partial F}{\partial y_i'} \delta y_i' \, dx \right\} \\
&= \sum_{i=1}^{n} \left\{ \int_{a}^{b} \frac{\partial F}{\partial y_i} \delta y_i \, dx - \int_{a}^{b} \left[\frac{d}{dx} \frac{\partial F}{\partial y_i'} \right] \delta y_i \, dx + \left[\frac{\partial F}{\partial y_i'} \delta y_i \right] \Big|_{a}^{b} \right\}
\end{aligned}$$

The boundary terms are zero because $\delta y_i(a) = \delta y_i(b) = 0$ for $i = 1, \ldots, n$. Recombining the integrals produces

$$\delta J = \sum_{i=1}^{n} \int_{a}^{b} \left[\frac{\partial F}{\partial y_i} - \frac{d}{dx} \frac{\partial F}{\partial y_i'} \right] \delta y_i \, dx$$

The necessary condition $\delta J = 0$ holds for all admissible variations $\delta \mathbf{y}$. Every integral must vanish because the admissible variations δy_i are independent. This fact is easily demonstrated by varying a particular δy_i and setting the remainder of the variations $\delta y_j = 0$. Hence, we extract the following n equations:

$$\int_{a}^{b} \left[\frac{\partial F}{\partial y_i} - \frac{d}{dx} \frac{\partial F}{\partial y_i'} \right] \delta y_i \, dx = 0, \qquad i = 1, \ldots, n$$

Applying the Fundamental Lemma 4.6.1 gives

$$\frac{\partial F}{\partial y_i} - \frac{d}{dx} \frac{\partial F}{\partial y_i'} = 0, \qquad i = 1, \ldots, n$$

A simpler proof is provided by applying lemma 4.6.3 to equation (8.1.4) with $\delta J = 0$. We immediately conclude that

$$\frac{d}{dx} \frac{\partial F}{\partial y_i'} = \frac{\partial F}{\partial y_i}, \qquad i = 1, \ldots, n$$

because the variations δy_i are independent. $\square$

The commutative property of equation (7.4.5) allows us to compute the 2nd variation according to

$$\delta^2 J = \delta^2 \int_{a}^{b} F \, dx = \int_{a}^{b} \delta^2 F \, dx$$

The 2nd variation $\delta^2 F$ is defined using the Hessian matrix. We define $\mathbf{u}$ and $\delta \mathbf{u}$ as follows:

$$\mathbf{u} = [y_1, \ldots, y_n, y_1', \ldots, y_n', x]^T, \qquad y_i = u_i, \quad y_i' = u_{n+i}, \quad x = u_{2n+1}$$
$$\delta \mathbf{u} = [\delta y_1, \ldots, \delta y_n, \delta y_1', \ldots, \delta y_n', \delta x]^T, \qquad \delta y_i = \delta u_i, \quad \delta y_i' = \delta u_{n+i}, \quad \delta x = \delta u_{2n+1}$$

The Hessian matrix is

$$\mathbf{H}_{ij}(\mathbf{u}) = \frac{\partial^2}{\partial u_i \partial u_j} F(\mathbf{u})$$

and $\delta^2 F$ is computed as

$$\delta^2 F \equiv \frac{1}{2} (\delta \mathbf{u})^T \mathbf{H}(\mathbf{u}) \, \delta \mathbf{u}$$

Further algebraic manipulation of $\delta^2 F$ is required to formulate $\delta^2 J$. Recall that $\delta x = 0$. Performing the matrix multiplication results in

$$\delta^2 F \equiv \frac{1}{2} (\delta \mathbf{u})^T \mathbf{H}(\mathbf{u}) \, \delta \mathbf{u} = \frac{1}{2} \sum_{i=1}^{2n} \sum_{j=1}^{2n} \frac{\partial^2 F}{\partial u_i \partial u_j} \delta u_i \, \delta u_j$$

We wish to separate the δy_i terms from the $\delta y_i'$ terms.

$$\begin{aligned}
\delta^2 F &= \frac{1}{2} \sum_{i=1}^{n} \sum_{j=1}^{2n} \frac{\partial^2 F}{\partial u_i \partial u_j} \delta u_i \, \delta u_j + \frac{1}{2} \sum_{i=n+1}^{2n} \sum_{j=1}^{2n} \frac{\partial^2 F}{\partial u_i \partial u_j} \delta u_i \, \delta u_j \\
&= \frac{1}{2} \sum_{i=1}^{n} \left[\sum_{j=1}^{n} \frac{\partial^2 F}{\partial u_i \partial u_j} \delta u_i \, \delta u_j + \sum_{j=n+1}^{2n} \frac{\partial^2 F}{\partial u_i \partial u_j} \delta u_i \, \delta u_j \right] \\
&\quad + \frac{1}{2} \sum_{i=n+1}^{2n} \left[\sum_{j=1}^{n} \frac{\partial^2 F}{\partial u_i \partial u_j} \delta u_i \, \delta u_j + \sum_{j=n+1}^{2n} \frac{\partial^2 F}{\partial u_i \partial u_j} \delta u_i \, \delta u_j \right] \\
&= \frac{1}{2} \sum_{i=1}^{n} \sum_{j=1}^{n} \frac{\partial^2 F}{\partial u_i \partial u_j} \delta u_i \, \delta u_j + \frac{1}{2} \sum_{i=1}^{n} \sum_{j=n+1}^{2n} \frac{\partial^2 F}{\partial u_i \partial u_j} \delta u_i \, \delta u_j \\
&\quad + \frac{1}{2} \sum_{i=n+1}^{2n} \sum_{j=1}^{n} \frac{\partial^2 F}{\partial u_i \partial u_j} \delta u_i \, \delta u_j + \frac{1}{2} \sum_{i=n+1}^{2n} \sum_{j=n+1}^{2n} \frac{\partial^2 F}{\partial u_i \partial u_j} \delta u_i \, \delta u_j
\end{aligned}$$

Changing back to the function variables gives

$$\begin{aligned}
\delta^2 F &= \frac{1}{2} \sum_{i=1}^{n} \sum_{j=1}^{n} \frac{\partial^2 F}{\partial y_i \partial y_j} \delta y_i \, \delta y_j + \frac{1}{2} \sum_{i=1}^{n} \sum_{j=1}^{n} \frac{\partial^2 F}{\partial y_i \partial y_j'} \delta y_i \, \delta y_j' \\
&\quad + \frac{1}{2} \sum_{i=1}^{n} \sum_{j=1}^{n} \frac{\partial^2 F}{\partial y_i' \partial y_j} \delta y_i' \, \delta y_j + \frac{1}{2} \sum_{i=1}^{n} \sum_{j=1}^{n} \frac{\partial^2 F}{\partial y_i' \partial y_j'} \delta y_i' \, \delta y_j'
\end{aligned}$$

The symmetry of the Hessian matrix gives our final result for $\delta^2 F$.

$$\delta^2 F = \frac{1}{2} \sum_{i=1}^{n} \sum_{j=1}^{n} \left[\frac{\partial^2 F}{\partial y_i \partial y_j} \delta y_i \, \delta y_j + 2 \frac{\partial^2 F}{\partial y_i \partial y_j'} \delta y_i \, \delta y_j' + \frac{\partial^2 F}{\partial y_i' \partial y_j'} \delta y_i' \, \delta y_j' \right]$$

8.1. FIXED END POINTS

The 2nd variation $\delta^2 J$ can now be computed.

$$\delta^2 J = \frac{1}{2} \int_a^b \sum_{i=1}^n \sum_{j=1}^n \left[\frac{\partial^2 F}{\partial y_i \partial y_j} \delta y_i \, \delta y_j + 2 \frac{\partial^2 F}{\partial y_i \partial y'_j} \delta y_i \, \delta y'_j + \frac{\partial^2 F}{\partial y'_i \partial y'_j} \delta y'_i \, \delta y'_j \right] dx$$

We note the following anti-derivative:

$$\int [\delta y'_i \, \delta y_j + \delta y_i \, \delta y'_j] dx = \delta y_i \, \delta y_j$$

This result is required in order to transform the middle term in brackets of $\delta^2 J$ using integration by parts.

$$\int_a^b \frac{\partial^2 F}{\partial y_i \partial y'_j} [\delta y'_i \, \delta y_j + \delta y_i \, \delta y'_j] dx = -\int_a^b \left[\frac{d}{dx} \frac{\partial^2 F}{\partial y_i \partial y'_j} \right] \delta y_i \, \delta y_j \, dx + \left[\frac{\partial^2 F}{\partial y_i \partial y'_j} \delta y_i \, \delta y_j \right]\Big|_a^b$$

The boundary term vanishes because the variations vanish at the endpoints.

$$\int_a^b \frac{\partial^2 F}{\partial y_i \partial y'_j} [\delta y'_i \, \delta y_j + \delta y_i \, \delta y'_j] dx = -\int_a^b \left[\frac{d}{dx} \frac{\partial^2 F}{\partial y_i \partial y'_j} \right] \delta y_i \, \delta y_j \, dx$$

The following double sum is transformed using the expression above.

$$\int_a^b \left[\sum_{i=1}^n \sum_{j=1}^n 2 \frac{\partial^2 F}{\partial y_i \partial y'_j} \delta y_i \, \delta y'_j \right] dx = \int_a^b \sum_{i=1}^n \sum_{j=1}^n \frac{\partial^2 F}{\partial y_i \partial y'_j} [\delta y'_i \, \delta y_j + \delta y_i \, \delta y'_j] dx$$

$$= -\int_a^b \sum_{i=1}^n \sum_{j=1}^n \left[\frac{d}{dx} \frac{\partial^2 F}{\partial y_i \partial y'_j} \right] \delta y_i \, \delta y_j \, dx$$

The manipulation in the first line is possible because the Hessian is symmetric. $\delta^2 J$ can now be written as

$$\delta^2 J = \frac{1}{2} \int_a^b \sum_{i=1}^n \sum_{j=1}^n \left[\left(\frac{\partial^2 F}{\partial y_i \partial y_j} - \frac{d}{dx} \frac{\partial^2 F}{\partial y_i \partial y'_j} \right) \delta y_i \, \delta y_j + \frac{\partial^2 F}{\partial y'_i \partial y'_j} \delta y'_i \, \delta y'_j \right] dx$$

Using matrix and vector notation, we write

$$\delta^2 J = \int_a^b \left[(\delta \mathbf{y}')^T \mathbf{P}(x)\, \delta \mathbf{y}' + (\delta \mathbf{y})^T \mathbf{Q}(x)\, \delta \mathbf{y} \right] dx \qquad (8.1.7)$$

$$\mathbf{P}_{ij}(x) = \frac{1}{2} \frac{\partial^2 F}{\partial y_i' \partial y_j'}$$

$$\mathbf{Q}_{ij}(x) = \frac{1}{2} \left(\frac{\partial^2 F}{\partial y_i \partial y_j} - \frac{d}{dx} \frac{\partial^2 F}{\partial y_i \partial y_j'} \right)$$

The Legendre necessary condition is provided by the following theorem. The proof is omitted due to its complexity.

Theorem 8.1.2 (Legendre Condition). *Consider the closed interval $I = [a, b]$, the set of functions*

$$\mathcal{U} = \left\{ \mathbf{y} : I \to \mathbb{R}^n \;\middle|\; \mathbf{y}(a) = \mathbf{y}_a,\ \mathbf{y}(b) = \mathbf{y}_b,\ y_i \in \mathcal{C}^1(I),\ i = 1, \ldots, n \right\}$$

and the functional $J : \mathcal{U} \to \mathbb{R}$ defined by

$$J[\mathbf{y}] = \int_a^b F(x, \mathbf{y}, \mathbf{y}')\, dx \qquad (8.1.8)$$

where F has continuous partial derivatives up to order 3. A necessary condition for a local minimum of J over $\mathcal{U}$ is that $\mathbf{P}(x)$ be positive semi-definite along the solution curve $\mathbf{y}(x)$ for $x \in I$.

Positive semi-definite means that

$$(\delta \mathbf{y}')^T \mathbf{P}(x)\, \delta \mathbf{y}' \geq 0, \qquad \forall \delta \mathbf{y}' \neq \mathbf{0}$$

This page intentionally left blank.

8.2 Isoparametric Conditions

Consider the minimization of the functional

$$J[\mathbf{y}] = \int_a^b F(x, \mathbf{y}, \mathbf{y}') \, dx \tag{8.2.1}$$

where $\mathbf{y}(x)$ has fixed endpoints $\mathbf{y}(a) = \mathbf{y}_a$ and $\mathbf{y}(b) = \mathbf{y}_b$. The minimization problem also satisfies the condition

$$K[\mathbf{y}] = \int_a^b G(x, \mathbf{y}, \mathbf{y}') \, dx = C \tag{8.2.2}$$

where C is a constant. F and G have continuous partial derivatives up to order 2. Set $I = [a, b]$. The set of possible solutions $\mathcal{U}$ is defined as

$$\mathcal{U} = \left\{ \mathbf{y} : I \to \mathbb{R}^n \,\middle|\, \mathbf{y}(a) = \mathbf{y}_a,\ \mathbf{y}(b) = \mathbf{y}_b,\ y_i \in \mathcal{C}^1(I),\ i = 1, \ldots, n \right\}$$

The admissible variations are

$$\mathcal{V} = \left\{ \delta \mathbf{y} : I \to \mathbb{R}^n \,\middle|\, \delta y_i \in \mathcal{C}_0^1(I),\ i = 1, \ldots, n \right\}$$

The minimization problem is solved using the method of Lagrange multipliers. The variation of the functional K is zero because the variation of the constant C is zero.

$$\delta K = \int_a^b \delta G \, dx = 0$$

We multiply δK by an unspecified constant λ and add the result to the variation δJ. In order words, we add zero to δJ. Integration by parts yields

$$\begin{aligned}
\delta J &= \delta J + \lambda \delta K \\
&= \int_a^b \delta F \, dx + \lambda \int_a^b \delta G \, dx \\
&= \int_a^b \sum_{i=1}^n \left[\frac{\partial F}{\partial y_i} \delta y_i + \frac{\partial F}{\partial y_i'} \delta y_i' \right] dx + \lambda \int_a^b \sum_{i=1}^n \left[\frac{\partial G}{\partial y_i} \delta y_i + \frac{\partial G}{\partial y_i'} \delta y_i' \right] dx \\
&= \sum_{i=1}^n \int_a^b \left[\frac{\partial F}{\partial y_i} \delta y_i + \frac{\partial F}{\partial y_i'} \delta y_i' \right] dx + \lambda \sum_{i=1}^n \int_a^b \left[\frac{\partial G}{\partial y_i} \delta y_i + \frac{\partial G}{\partial y_i'} \delta y_i' \right] dx \\
&= \sum_{i=1}^n \int_a^b \left[\frac{\partial F}{\partial y_i} - \frac{d}{dx} \frac{\partial F}{\partial y_i'} \right] \delta y_i \, dx + \sum_{i=1}^n \left[\frac{\partial F}{\partial y_i'} \delta y_i \right] \bigg|_a^b \\
&\quad + \lambda \sum_{i=1}^n \int_a^b \left[\frac{\partial G}{\partial y_i} - \frac{d}{dx} \frac{\partial G}{\partial y_i'} \right] \delta y_i \, dx + \lambda \sum_{i=1}^n \left[\frac{\partial G}{\partial y_i'} \delta y_i \right] \bigg|_a^b
\end{aligned}$$

8.2. ISOPARAMETRIC CONDITIONS

The boundary terms vanish because there is no variation at the endpoints. Combining the remaining integrals gives

$$\delta J = \sum_{i=1}^{n} \int_{a}^{b} \left[\frac{\partial F}{\partial y_i} - \frac{d}{dx} \frac{\partial F}{\partial y'_i} + \lambda \left(\frac{\partial G}{\partial y_i} - \frac{d}{dx} \frac{\partial G}{\partial y'_i} \right) \right] \delta y_i \, dx$$

The 1st order necessary condition requires $\delta J = 0$ for all admissible variations. The admissible variations being independent means that we can vary any particular δy_i and set the remainder of the variations $\delta y_j = 0$. Hence, each integral must vanish.

$$\int_{a}^{b} \left[\frac{\partial F}{\partial y_i} - \frac{d}{dx} \frac{\partial F}{\partial y'_i} + \lambda \left(\frac{\partial G}{\partial y_i} - \frac{d}{dx} \frac{\partial G}{\partial y'_i} \right) \right] \delta y_i \, dx = 0, \qquad i = 1, \ldots, n$$

Application of the Fundamental Lemma 4.6.1 requires that the term in brackets is zero.

$$\frac{\partial F}{\partial y_i} - \frac{d}{dx} \frac{\partial F}{\partial y'_i} + \lambda \left(\frac{\partial G}{\partial y_i} - \frac{d}{dx} \frac{\partial G}{\partial y'_i} \right) = 0, \qquad i = 1, \ldots, n$$

Theorem 8.2.1. *Consider the closed interval $I = [a, b]$, the set of functions*

$$\mathcal{U} = \left\{ \mathbf{y} : I \to \mathbb{R}^n \,\middle|\, \mathbf{y}(a) = \mathbf{y}_a, \, \mathbf{y}(b) = \mathbf{y}_b, \, y_i \in C^1(I), \, i = 1, \ldots, n \right\}$$

and the functional $J : \mathcal{U} \to \mathbb{R}$ defined by

$$J[\mathbf{y}] = \int_{a}^{b} F(x, \mathbf{y}, \mathbf{y}') \, dx \qquad (8.2.3)$$

where F has continuous partial derivatives up to order 2. Let G have continuous partial derivatives up to order 2. A necessary condition for a local minimum of J over $\mathcal{U}$ subject to the isoparametric condition

$$K[\mathbf{y}] = \int_{a}^{b} G(x, \mathbf{y}, \mathbf{y}') \, dx = C \qquad (8.2.4)$$

is that the following differential equations be satisfied:

$$\frac{\partial F}{\partial y_i} - \frac{d}{dx} \frac{\partial F}{\partial y'_i} + \lambda \left(\frac{\partial G}{\partial y_i} - \frac{d}{dx} \frac{\partial G}{\partial y'_i} \right) = 0, \qquad i = 1, \ldots, n \qquad (8.2.5)$$

The solution for λ is obtained while solving equation (8.2.5) in conjunction with condition (8.2.4). Equation (8.2.5) is equivalent to the Euler differential equation for the augmented functional

$$\tilde{J}[\mathbf{y}] = J[\mathbf{y}] + \lambda K[\mathbf{y}] = \int_{a}^{b} \tilde{F}(x, \mathbf{y}, \mathbf{y}') \, dx \qquad (8.2.6)$$

$$\tilde{F}(x, \mathbf{y}, \mathbf{y}') = F(x, \mathbf{y}, \mathbf{y}') + \lambda G(x, \mathbf{y}, \mathbf{y}') \qquad (8.2.7)$$

$$\frac{\partial \tilde{F}}{\partial y_i} - \frac{d}{dx} \frac{\partial \tilde{F}}{\partial y'_i} = 0, \qquad i = 1, \ldots, n \qquad (8.2.8)$$

8.3 Auxiliary Constraints

We again wish to the minimize the functional

$$J[\mathbf{y}] = \int_a^b F(x, \mathbf{y}, \mathbf{y}') \, dx \tag{8.3.1}$$

where $\mathbf{y}(x)$ has fixed endpoints $\mathbf{y}(a) = \mathbf{y}_a$ and $\mathbf{y}(b) = \mathbf{y}_b$. The following auxiliary constraints must also be satisfied:

$$G_k(x, \mathbf{y}) = 0, \qquad \forall x \in [a, b], \quad k = 1, \ldots, m \tag{8.3.2}$$

where $m < n$. F and G_k have continuous partial derivatives up to order 2. The set of possible solutions $\mathcal{U}$ is again defined as

$$\mathcal{U} = \left\{ \mathbf{y} : I \to \mathbb{R}^n \,\big|\, \mathbf{y}(a) = \mathbf{y}_a, \mathbf{y}(b) = \mathbf{y}_b, \quad y_i \in \mathcal{C}^1(I), \quad i = 1, \ldots, n \right\}$$

where $I = [a, b]$. The admissible variations are

$$\mathcal{V} = \left\{ \delta \mathbf{y} : I \to \mathbb{R}^n \,\big|\, \delta y_i \in \mathcal{C}_0^1(I), \quad i = 1, \ldots, n \right\}$$

We must caution that the variations δy_i are not independent and cannot be freely varied. However, the method of Lagrange multipliers can be employed to solve the problem.

We begin by computing the variation of G_k in equation (8.3.2) for $k = 1, \ldots, m$.

$$\delta G_k = \sum_{i=1}^n \frac{\partial G_k}{\partial y_i} \delta y_i = 0, \qquad k = 1, \ldots, m \tag{8.3.3}$$

Next, we multiply each equation by an unknown function $\lambda_k(x)$ and sum the equations

$$\sum_{k=1}^m \lambda_k \, \delta G_k = \sum_{k=1}^m \sum_{i=1}^n \lambda_k \frac{\partial G_k}{\partial y_i} \delta y_i = 0$$

Integrating this result from a to b, interchanging the order of summation, and interchanging summation and integration results in

$$\begin{aligned}
\int_a^b \sum_{k=1}^m \lambda_k \, \delta G_k \, dx &= \int_a^b \sum_{k=1}^m \sum_{i=1}^n \lambda_k \frac{\partial G_k}{\partial y_i} \delta y_i \, dx \\
&= \int_a^b \sum_{i=1}^n \sum_{k=1}^m \lambda_k \frac{\partial G_k}{\partial y_i} \delta y_i \, dx \\
&= \sum_{i=1}^n \int_a^b \sum_{k=1}^m \lambda_k \frac{\partial G_k}{\partial y_i} \delta y_i \, dx
\end{aligned}$$

8.3. AUXILIARY CONSTRAINTS

Recall that these equations are equal to zero.

$$\int_a^b \sum_{k=1}^m \lambda_k \, \delta G_k \, dx = \sum_{i=1}^n \int_a^b \sum_{k=1}^m \lambda_k \frac{\partial G_k}{\partial y_i} \delta y_i \, dx = 0 \qquad (8.3.4)$$

Now add equation (8.3.4) to the variation of J. We also note that $\delta J = 0$ is the 1st order necessary condition. Thus we add zero in the following manner and integrate by parts:

$$\delta J + \int_a^b \sum_{k=1}^m \lambda_k \, \delta G_k \, dx = 0$$

$$\sum_{i=1}^n \int_a^b \left[\frac{\partial F}{\partial y_i} \delta y_i + \frac{\partial F}{\partial y_i'} \delta y_i' \right] dx + \sum_{i=1}^n \int_a^b \sum_{k=1}^m \lambda_k \frac{\partial G_k}{\partial y_i} \delta y_i \, dx = 0$$

$$\sum_{i=1}^n \int_a^b \left[\frac{\partial F}{\partial y_i} - \frac{d}{dx} \frac{\partial F}{\partial y_i'} \right] \delta y_i \, dx + \sum_{i=1}^n \left[\frac{\partial F}{\partial y_i'} \delta y_i \right]_a^b$$

$$+ \sum_{i=1}^n \int_a^b \sum_{k=1}^m \lambda_k \frac{\partial G_k}{\partial y_i} \delta y_i \, dx = 0$$

The boundary terms vanish because there is no variation at the endpoints. Combining the integrals gives

$$\sum_{i=1}^n \int_a^b \left[\frac{\partial F}{\partial y_i} - \frac{d}{dx} \frac{\partial F}{\partial y_i'} + \sum_{k=1}^m \lambda_k \frac{\partial G_k}{\partial y_i} \right] \delta y_i \, dx = 0 \qquad (8.3.5)$$

We cannot say the terms in brackets must vanish because the variations δy_i are not independent but related by equation (8.3.3).

Now consider the first m variables $y_1, \ldots, y_m$ to be dependent and the remaining $y_{m+1}, \ldots, y_n$ to be independent. The sum in equation (8.3.5) is partitioned in the follow way:

$$\sum_{i=1}^m \int_a^b \left[\frac{\partial F}{\partial y_i} - \frac{d}{dx} \frac{\partial F}{\partial y_i'} + \sum_{k=1}^m \lambda_k \frac{\partial G_k}{\partial y_i} \right] \delta y_i \, dx$$

$$+ \sum_{i=m+1}^n \int_a^b \left[\frac{\partial F}{\partial y_i} - \frac{d}{dx} \frac{\partial F}{\partial y_i'} + \sum_{k=1}^m \lambda_k \frac{\partial G_k}{\partial y_i} \right] \delta y_i \, dx = 0 \qquad (8.3.6)$$

We can choose λ_k in a manner which causes each bracket in the first sum of equation (8.3.6) to vanish.

$$\frac{\partial F}{\partial y_i} - \frac{d}{dx}\frac{\partial F}{\partial y'_i} + \sum_{k=1}^{m} \lambda_k \frac{\partial G_k}{\partial y_i} = 0, \qquad \forall x \in I, \quad i = 1, \ldots, m$$

These are m equations in the m unknown functions λ_k. The vector and matrix form of the equations is

$$A_{ik} = \frac{\partial G_k}{\partial y_i}, \qquad l_k = \lambda_k, \qquad b_i = -\left[\frac{\partial F}{\partial y_i} - \frac{d}{dx}\frac{\partial F}{\partial y'_i}\right]$$

$$\mathbf{A} = [A_{ik}], \qquad \mathbf{l} = [\lambda_1, \ldots, \lambda_m]^T, \qquad \mathbf{b} = [b_1, \ldots, b_m]^T$$

$$\mathbf{A}\mathbf{l} = \mathbf{b}$$

A unique solution for λ_k exists if $\det \mathbf{A} \neq 0$. Note that the matrix equation is solved for every $x \in I$ to determine $\lambda_k(x)$. We now have

$$\frac{\partial F}{\partial y_i} - \frac{d}{dx}\frac{\partial F}{\partial y'_i} + \sum_{k=1}^{m} \lambda_k \frac{\partial G_k}{\partial y_i} = 0, \qquad i = 1, \ldots, m$$

The remaining sum in equation (8.3.6) contains independent variables and thus δy_i can be arbitrarily varied. This implies that

$$\int_a^b \left[\frac{\partial F}{\partial y_i} - \frac{d}{dx}\frac{\partial F}{\partial y'_i} + \sum_{k=1}^{m} \lambda_k \frac{\partial G_k}{\partial y_i}\right] \delta y_i \, dx, \qquad i = m+1, \ldots, n$$

Application of the Fundamental Lemma 4.6.1 gives

$$\frac{\partial F}{\partial y_i} - \frac{d}{dx}\frac{\partial F}{\partial y'_i} + \sum_{k=1}^{m} \lambda_k \frac{\partial G_k}{\partial y_i} = 0, \qquad i = m+1, \ldots, n$$

Combining the results above, we finally have

$$\frac{\partial F}{\partial y_i} - \frac{d}{dx}\frac{\partial F}{\partial y'_i} + \sum_{k=1}^{m} \lambda_k \frac{\partial G_k}{\partial y_i} = 0, \qquad i = 1, \ldots, n$$

8.3. AUXILIARY CONSTRAINTS

Theorem 8.3.1. *Consider the closed interval $I = [a, b]$, the set of functions*

$$\mathcal{U} = \left\{ \mathbf{y} : I \to \mathbb{R}^n \mid \mathbf{y}(a) = \mathbf{y}_a, \, \mathbf{y}(b) = \mathbf{y}_b, \quad y_i \in \mathcal{C}^1(I), \quad i = 1, \ldots, n \right\}$$

and the functional $J : \mathcal{U} \to \mathbb{R}$ defined by

$$J[\mathbf{y}] = \int_a^b F(x, \mathbf{y}, \mathbf{y}') \, dx \tag{8.3.7}$$

where F has continuous partial derivatives up to order 2. Let G_k have continuous partial derivatives up to order 2. A necessary condition for a local minimum of J over $\mathcal{U}$ subject to the auxiliary conditions

$$G_k(x, \mathbf{y}) = 0, \qquad \forall x \in I, \quad k = 1, \ldots, m \tag{8.3.8}$$

is that the following differential equations be satisfied:

$$\frac{\partial F}{\partial y_i} - \frac{d}{dx} \frac{\partial F}{\partial y_i'} + \sum_{k=1}^m \lambda_k \frac{\partial G_k}{\partial y_i} = 0, \quad i = 1, \ldots, n \tag{8.3.9}$$

The solution for the λ_k are obtained while solving equation (8.3.9) in conjunction with the equations (8.3.8). Equation (8.3.9) is equivalent to the Euler differential equation for the augmented functional using the augmented vector $\mathbf{u}$.

$$\mathbf{u} = [y_1, \ldots, y_n, \lambda_1, \ldots, \lambda_m]^T = [\mathbf{y}^T, \mathbf{1}^T]^T$$

$$\tilde{J}[\mathbf{u}] = \int_a^b \left[F(x, \mathbf{y}, \mathbf{y}') + \sum_{k=1}^m \lambda_k G_k(x, \mathbf{y}) \right] dx \tag{8.3.10}$$

$$\tilde{F}(\mathbf{u}) = F(x, \mathbf{y}, \mathbf{y}') + \sum_{k=1}^m \lambda_k G_k(x, \mathbf{y}) \tag{8.3.11}$$

$$\frac{\partial \tilde{F}}{\partial u_i} - \frac{d}{dx} \frac{\partial \tilde{F}}{\partial u_i'} = 0, \quad i = 1, \ldots, n + m \tag{8.3.12}$$

8.4 Geodesics

Suppose that we wish to compute the shortest path between two points on a surface. The path is defined parametrically by the points $(x(t), y(t), z(t))$ for $t \in [0, 1]$. The end points are $(x(0), y(0), z(0)) = (x_0, y_0, z_0)$ and $(x(1), y(1), z(1)) = (x_1, y_1, z_1)$. The path is constrained to a surface implicitly defined by $G(x, y, z) = 0$. Such paths are called *geodesics* and are the solution to the following variational problem. Let J be the functional which computes arc length.

$$J[x, y, z] = \int_0^1 F(x', y', z') \, dx = \int_0^1 \sqrt{(x')^2 + (y')^2 + (z')^2} \, dt \qquad (8.4.1)$$

The minimization of the functional is subject to the condition $G(x, y, z) = 0$.

Applying the results of the previous section, the shortest curve is obtained by solving the following set of equations:

$$\frac{\partial F}{\partial x} - \frac{d}{dt}\frac{\partial F}{\partial x'} + \lambda \frac{\partial G}{\partial x} = 0$$

$$\frac{\partial F}{\partial y} - \frac{d}{dt}\frac{\partial F}{\partial y'} + \lambda \frac{\partial G}{\partial y} = 0$$

$$\frac{\partial F}{\partial z} - \frac{d}{dt}\frac{\partial F}{\partial z'} + \lambda \frac{\partial G}{\partial z} = 0$$

$$G(x, y, z) = 0$$

Remember that $\lambda \equiv \lambda(x)$ is an unknown function.

Computing the partial derivatives of F and multiplying through by -1 gives

$$\frac{d}{dt}\left[\frac{x'}{\sqrt{(x')^2 + (y')^2 + (z')^2}}\right] - \lambda \frac{\partial G}{\partial x} = 0 \qquad (8.4.2)$$

$$\frac{d}{dt}\left[\frac{y'}{\sqrt{(x')^2 + (y')^2 + (z')^2}}\right] - \lambda \frac{\partial G}{\partial y} = 0 \qquad (8.4.3)$$

$$\frac{d}{dt}\left[\frac{z'}{\sqrt{(x')^2 + (y')^2 + (z')^2}}\right] - \lambda \frac{\partial G}{\partial z} = 0 \qquad (8.4.4)$$

$$G(x, y, z) = 0 \qquad (8.4.5)$$

8.4. GEODESICS

We can show that the great circle distance is the shortest path between two points on the surface of a sphere. Instead of solving equations (8.4.2)-(8.4.5) directly, we show that the known solution satisfies these equations. Let a sphere of radius R be centered at the origin and consider two points (x_0, y_0, z_0) and (x_1, y_1, z_1) on the surface. We further simplify the problem by choosing a coordinate system where the two points lie in the same plane. Let $\mathbf{r}_0 = \langle x_0, y_0, z_0 \rangle$ and $\mathbf{r}_1 = \langle x_1, y_1, z_1 \rangle$ be the vectors connecting the origin to these points. Remember that $\|\mathbf{r}_0\|_2 = \|\mathbf{r}_1\| = R$. Choose $\mathbf{r}_0$ to be the direction of the new x-axis. The new z-axis is in the direction of $\mathbf{z} = \mathbf{r}_0 \times \mathbf{r}_1$. Then the new y-axis is in the direction of $\mathbf{z} \times \mathbf{r}_0$.

The great circle is defined by the following equations:

$$x(t) = R \cos(t)$$
$$y(t) = R \sin(t)$$

where $0 \leq t \leq \theta$. The final angle θ is defined by the dot product of $\mathbf{r}_0$ and $\mathbf{r}_1$.

$$\mathbf{r}_0 \cdot \mathbf{r}_1 = R^2 \cos \theta$$
$$\cos \theta = \frac{\mathbf{r}_0 \cdot \mathbf{r}_1}{R^2}$$
$$\theta = \cos^{-1}\left(\frac{\mathbf{r}_0 \cdot \mathbf{r}_1}{R^2}\right)$$

This definition ensures that we obtain the shortest path on a great circle. In other words, the path where $0 \leq \theta \leq \pi$.

Our solution must satisfy the set of equations

$$\frac{d}{dt}\left[\frac{x'}{\sqrt{(x')^2 + (y')^2}}\right] - 2\lambda x = 0$$

$$\frac{d}{dt}\left[\frac{y'}{\sqrt{(x')^2 + (y')^2}}\right] - 2\lambda y = 0$$

$$x^2 + y^2 - R^2 = 0$$

Substituting the functions $x(t)$ and $y(t)$ gives

$$\frac{d}{dt}[-\sin(t)] - 2\lambda R \cos(t) = 0$$

$$\frac{d}{dt}[\cos(t)] - 2\lambda R \sin(t) = 0$$

$$R^2 \cos^2(t) + R^2 \sin^2(t) - R^2 = 0$$

The first two equations are clearly satisfied with $\lambda = -1/(2R)$. The third equation is a fundamental trigonometric identity.

8.5 Variable End Points

We now consider the more general functional

$$J[\mathbf{y}] = \int_a^b F(x, \mathbf{y}, \mathbf{y}') \, dx + f(\mathbf{y}(b)) - g(\mathbf{y}(a)) \tag{8.5.1}$$

where F has continuous partial derivatives up to order 2 in each of its arguments. f and g map $\mathbb{R}^n$ to $\mathbb{R}$ and $f, g \in \mathcal{C}^1(\mathbb{R}^n)$. We no longer constrain the values of $\mathbf{y}$ at the endpoints a and b. Let $I = [a, b]$ as usual. The set of possible solutions $\mathcal{U}$ now becomes

$$\mathcal{U} = \left\{ \mathbf{y} : I \to \mathbb{R}^n \ \big| \ y_i \in \mathcal{C}^1(I), \quad i = 1, \ldots, n \right\}$$

The local minimization problem is

$$\mathbf{y}^* = \arg \operatorname*{local\,min}_{\mathbf{y} \in \mathcal{U}} J[\mathbf{y}] \tag{8.5.2}$$

We proceed to develop the necessary equations for a local minimum. The space of admissible variations is identical to the space of solutions: $\mathcal{V} = \mathcal{U}$. Using vertical bar notation to denote the location where $\mathbf{y}$ and its variation are evaluated, the 1st variation is

$$\begin{aligned}
\delta J &= \int_a^b \delta F \, dx + \delta f - \delta g \\
&= \sum_{i=1}^n \int_a^b \left[\frac{\partial F}{\partial y_i} \delta y_i + \frac{\partial F}{\partial y_i'} \delta y_i' \right] dx + \sum_{i=1}^n \left[\frac{\partial f}{\partial y_i} \delta y_i \right]\bigg|_b - \sum_{i=1}^n \left[\frac{\partial g}{\partial y_i} \delta y_i \right]\bigg|_a \\
&= \sum_{i=1}^n \left\{ \int_a^b \frac{\partial F}{\partial y_i} \delta y_i \, dx + \int_a^b \frac{\partial F}{\partial y_i'} \delta y_i' \, dx \right\} \\
&\quad + \sum_{i=1}^n \left[\frac{\partial f}{\partial y_i} \delta y_i \right]\bigg|_b - \sum_{i=1}^n \left[\frac{\partial g}{\partial y_i} \delta y_i \right]\bigg|_a \\
&= \sum_{i=1}^n \left\{ \int_a^b \frac{\partial F}{\partial y_i} \delta y_i \, dx - \int_a^b \left[\frac{d}{dx} \frac{\partial F}{\partial y_i'} \right] \delta y_i \, dx + \left[\frac{\partial F}{\partial y_i'} \delta y_i \right]\bigg|_a^b \right\} \\
&\quad + \sum_{i=1}^n \left[\frac{\partial f}{\partial y_i} \delta y_i \right]\bigg|_b - \sum_{i=1}^n \left[\frac{\partial g}{\partial y_i} \delta y_i \right]\bigg|_a \\
&= \sum_{i=1}^n \int_a^b \left[\frac{\partial F}{\partial y_i} - \frac{d}{dx} \frac{\partial F}{\partial y_i'} \right] \delta y_i \, dx + \sum_{i=1}^n \left[\left(\frac{\partial F}{\partial y_i'} + \frac{\partial f}{\partial y_i} \right) \delta y_i \right]\bigg|_b \\
&\quad - \sum_{i=1}^n \left[\left(\frac{\partial F}{\partial y_i'} + \frac{\partial g}{\partial y_i} \right) \delta y_i \right]\bigg|_a
\end{aligned}$$

8.5. VARIABLE END POINTS

The variation δJ is written as

$$\delta J = \delta J_1 + \phi(b) - \psi(a) \tag{8.5.3}$$

$$\delta J_1 = \sum_{i=1}^{n} \int_a^b \left[\frac{\partial F}{\partial y_i} - \frac{d}{dx} \frac{\partial F}{\partial y_i'} \right] \delta y_i \, dx$$

$$\phi(b) = \sum_{i=1}^{n} \left[\left(\frac{\partial F}{\partial y_i'} + \frac{\partial f}{\partial y_i} \right) \delta y_i \right]_b$$

$$\psi(a) = \sum_{i=1}^{n} \left[\left(\frac{\partial F}{\partial y_i'} + \frac{\partial g}{\partial y_i} \right) \delta y_i \right]_a$$

If the variations vanish at the endpoints, then $\phi(b) = \psi(a) = 0$. The condition $\delta J = 0$ implies $\delta J_1 = 0$ and we recover the differential equations (8.1.6). Hence the term in brackets for δJ_1 is zero. If the variations at the endpoints do not vanish, then we must have

$$\left[\frac{\partial F}{\partial y_i'} + \frac{\partial f}{\partial y_i} \right]_b = 0, \quad \left[\frac{\partial F}{\partial y_i'} + \frac{\partial g}{\partial y_i} \right]_a = 0, \quad i = 1, \ldots, n$$

This proves the following theorem.

Theorem 8.5.1. *Consider the closed interval* $I = [a, b]$, *the set of functions*

$$\mathcal{U} = \left\{ \mathbf{y} : I \to \mathbb{R}^n \mid y_i \in \mathcal{C}^1(I), \quad i = 1, \ldots, n \right\}$$

and the functional $J : \mathcal{U} \to \mathbb{R}$ *defined by*

$$J[\mathbf{y}] = \int_a^b F(x, \mathbf{y}, \mathbf{y}') \, dx + f(\mathbf{y}(b)) - g(\mathbf{y}(a)) \tag{8.5.4}$$

where F has continuous partial derivatives up to order 2 and $f, g \in \mathcal{C}^1(\mathbb{R}^n)$. A necessary condition for a local minimum of J over $\mathcal{U}$ is that the following equations be satisfied:

$$\frac{\partial F}{\partial y_i} - \frac{d}{dx} \frac{\partial F}{\partial y_i'} = 0, \quad i = 1, \ldots, n \tag{8.5.5}$$

$$\left[\frac{\partial F}{\partial y_i'} + \frac{\partial f}{\partial y_i} \right]_b = 0, \quad \left[\frac{\partial F}{\partial y_i'} + \frac{\partial g}{\partial y_i} \right]_a = 0, \quad i = 1, \ldots, n \tag{8.5.6}$$

For completeness, the 2nd variation is computed using

$$\delta^2 J = \int_a^b \delta^2 F \, dx + \delta^2 f - \delta^2 g$$

We now assume that $f, g, h, p \in \mathcal{C}^2(\mathbb{R}^n)$. This produces the equation

$$\begin{aligned}\delta^2 J &= \frac{1}{2}\int_a^b \sum_{i=1}^n \sum_{j=1}^n \left[\frac{\partial^2 F}{\partial y_i \partial y_j} \delta y_i \, \delta y_j + 2\frac{\partial^2 F}{\partial y_i \partial y_j'} \delta y_i \, \delta y_j' + \frac{\partial^2 F}{\partial y_i' \partial y_j'} \delta y_i' \, \delta y_j' \right] dx \\ &\quad + \frac{1}{2}\sum_{i=1}^n \sum_{j=1}^n \left[\frac{\partial^2 f}{\partial y_i \partial y_j} \delta y_i \, \delta y_j \right]\bigg|_b - \frac{1}{2}\sum_{i=1}^n \sum_{j=1}^n \left[\frac{\partial^2 g}{\partial y_i \partial y_j} \delta y_i \, \delta y_j \right]\bigg|_a \quad (8.5.7)\end{aligned}$$

Chapter 9

Functions with 1st and 2nd Derivatives

9.1 Fixed End Points

In this chapter, we investigate functionals containing multiple functions with 1st and 2nd derivatives. In particular, the following integral functions will be studied:

$$J[y_1, \ldots, y_n] = \int_a^b F(x, y_1, \ldots, y_n, y_1', \ldots, y_n', y_1'', \ldots, y_n'') \, dx \qquad (9.1.1)$$

F has continuous partial derivatives up to order 3 in each of its arguments. Vector notation significantly shortens the equations.

$$J[\mathbf{y}] = \int_a^b F(x, \mathbf{y}, \mathbf{y}', \mathbf{y}'') \, dx \qquad (9.1.2)$$

For the fixed end point problem, we must now specify the function and its derivative at the endpoints. The corresponding set of possible solutions $\mathcal{U}$ is

$$\mathcal{U} = \{\mathbf{y} : I \to \mathbb{R} \mid \mathbf{y}(a) = \mathbf{y}_a, \; \mathbf{y}(b) = \mathbf{y}_b, \; \mathbf{y}'(a) = \mathbf{y}_a', \; \mathbf{y}'(b) = \mathbf{y}_b', \\ y_i \in \mathcal{C}^2(I), \quad i = 1, \ldots, n\}$$

Our minimization problem is

$$\mathbf{y}^* = \arg \operatorname*{local\,min}_{\mathbf{y} \in \mathcal{U}} J[\mathbf{y}] \qquad (9.1.3)$$

No variation at the endpoints implies that the admissible variations are

$$\mathcal{V} = \{\mathbf{y} : I \to \mathbb{R} \mid \mathbf{y}(a) = \mathbf{y}(b) = \mathbf{y}'(a) = \mathbf{y}'(b) = \mathbf{0}, \\ y_i \in \mathcal{C}^2(I), \quad i = 1, \ldots, n\}$$

Let $F \equiv F(x, \mathbf{y}, \mathbf{y}', \mathbf{y}'')$ and compute δF using the simplified procedure. Integration by parts is applied multiple times.

$$\begin{aligned}
\delta J &= \int_a^b \delta F \, dx \\
&= \int_a^b \sum_{i=1}^n \left[\frac{\partial F}{\partial y_i} \delta y_i + \frac{\partial F}{\partial y_i'} \delta y_i' + \frac{\partial F_i}{\partial y_i''} \delta y_i'' \right] dx \\
&= \sum_{i=1}^n \int_a^b \frac{\partial F}{\partial y_i} \delta y_i \, dx + \sum_{i=1}^n \int_a^b \frac{\partial F}{\partial y_i'} \delta y_i' \, dx + \sum_{i=1}^n \int_a^b \frac{\partial F}{\partial y_i''} \delta y_i'' \, dx \\
&= \sum_{i=1}^n \int_a^b \frac{\partial F}{\partial y_i} \delta y_i \, dx - \sum_{i=1}^n \int_a^b \left[\frac{d}{dx} \frac{\partial F}{\partial y_i'} \right] \delta y_i \, dx + \sum_{i=1}^n \left[\frac{\partial F}{\partial y_i'} \delta y_i \right]\Big|_a^b \\
&\quad - \sum_{i=1}^n \int_a^b \left[\frac{d}{dx} \frac{\partial F}{\partial y_i''} \right] \delta y_i' \, dx + \sum_{i=1}^n \left[\frac{\partial F}{\partial y_i''} \delta y_i' \right]\Big|_a^b \\
&= \sum_{i=1}^n \int_a^b \frac{\partial F}{\partial y_i} \delta y_i \, dx - \sum_{i=1}^n \int_a^b \left[\frac{d}{dx} \frac{\partial F}{\partial y_i'} \right] \delta y_i \, dx + \sum_{i=1}^n \left[\frac{\partial F}{\partial y_i'} \delta y_i \right]\Big|_a^b \\
&\quad + \sum_{i=1}^n \int_a^b \left[\frac{d^2}{dx^2} \frac{\partial F}{\partial y_i''} \right] \delta y_i \, dx - \sum_{i=1}^n \left[\frac{d}{dx} \frac{\partial F}{\partial y_i''} \delta y_i \right]\Big|_a^b + \sum_{i=1}^n \left[\frac{\partial F}{\partial y_i''} \delta y_i' \right]\Big|_a^b \\
&= \sum_{i=1}^n \int_a^b \frac{\partial F}{\partial y_i} \delta y_i \, dx - \sum_{i=1}^n \int_a^b \left[\frac{d}{dx} \frac{\partial F}{\partial y_i'} \right] \delta y_i \, dx + \sum_{i=1}^n \int_a^b \left[\frac{d^2}{dx^2} \frac{\partial F}{\partial y_i''} \right] \delta y_i \, dx \\
&\quad + \sum_{i=1}^n \left[\frac{\partial F}{\partial y_i'} \delta y_i \right]\Big|_a^b - \sum_{i=1}^n \left[\frac{d}{dx} \frac{\partial F}{\partial y_i''} \delta y_i \right]\Big|_a^b + \sum_{i=1}^n \left[\frac{\partial F}{\partial y_i''} \delta y_i' \right]\Big|_a^b
\end{aligned}$$

Note that the boundary terms vanish. We combine the sums and the integrals to get

$$\delta J = \sum_{i=1}^n \int_a^b \left[\frac{\partial F}{\partial y_i} - \frac{d}{dx} \frac{\partial F}{\partial y_i'} + \frac{d^2}{dx^2} \frac{\partial F}{\partial y_i''} \right] \delta y_i \, dx$$

The 1st order necessary condition states that $\delta J = 0$. The variations are independent and thus each integral must vanish.

$$\int_a^b \left[\frac{\partial F}{\partial y_i} - \frac{d}{dx} \frac{\partial F}{\partial y_i'} + \frac{d^2}{dx^2} \frac{\partial F}{\partial y_i''} \right] \delta y_i \, dx = 0, \qquad i = 1, \ldots, n$$

Applying the Fundamental Lemma 4.6.1 gives

$$\frac{\partial F}{\partial y_i} - \frac{d}{dx} \frac{\partial F}{\partial y_i'} + \frac{d^2}{dx^2} \frac{\partial F}{\partial y_i''} = 0, \qquad i = 1, \ldots, n$$

9.1. FIXED END POINTS

These differential equations can be directly obtained by using lemma 4.6.5. Their solution is a necessary condition for a minimum of the functional J.

Theorem 9.1.1. *Consider the closed interval $I = [a, b]$, the set of functions*

$$\begin{aligned}\mathcal{U} &= \{\mathbf{y} : I \to \mathbb{R} \mid \mathbf{y}(a) = \mathbf{y}_a,\ \mathbf{y}(b) = \mathbf{y}_b,\ \mathbf{y}'(a) = \mathbf{y}'_a,\ \mathbf{y}'(b) = \mathbf{y}'_b, \\ & \quad y_i \in C^2(I),\quad i = 1, \ldots, n\}\end{aligned}$$

and the functional $J : \mathcal{U} \to \mathbb{R}$ defined by

$$J[y] = \int_a^b F(x, \mathbf{y}, \mathbf{y}', \mathbf{y}'')\, dx \qquad (9.1.4)$$

where F has continuous partial derivatives up to order 3. A necessary condition for a local minimum of J over $\mathcal{U}$ is that the following differential equations be satisfied:

$$\frac{\partial F}{\partial y_i} - \frac{d}{dx}\frac{\partial F}{\partial y'_i} + \frac{d^2}{dx^2}\frac{\partial F}{\partial y''_i} = 0, \qquad i = 1, \ldots, n \qquad (9.1.5)$$

The 2nd variation is computed with the equation

$$\delta^2 J = \int_a^b \delta^2 F\, dx$$

The computation of $\delta^2 F$ is performed in the following steps. First, define $\mathbf{u}$ and $\delta \mathbf{u}$.

$$\mathbf{u} = [y_1, \ldots, y_n, y'_1, \ldots, y'_n, y''_1, \ldots, y''_n, x]^T$$

$$y_i = u_i, \quad y'_i = u_{n+i}, \quad y''_i = u_{2n+i}, \quad x = u_{3n+1}$$

$$\delta \mathbf{u} = [\delta y_1, \ldots, \delta y_n, \delta y'_1, \ldots, \delta y'_n, \delta y''_1, \ldots, \delta y''_n, \delta x]^T$$

$$\delta y_i = \delta u_i, \quad \delta y'_i = \delta u_{n+i}, \quad \delta y''_i = \delta u_{2n+i}, \quad \delta x = \delta u_{3n+1}$$

Remember that $\delta x = 0$. The Hessian matrix is

$$\mathbf{H}_{ij}(\mathbf{u}) = \frac{\partial^2}{\partial u_i \partial u_j} F(\mathbf{u})$$

and $\delta^2 F$ is computed as

$$\delta^2 F \equiv \frac{1}{2}(\delta \mathbf{u})^T \mathbf{H}(\mathbf{u})\, \delta \mathbf{u} = \frac{1}{2}\sum_{i=1}^{3n}\sum_{j=1}^{3n} \frac{\partial^2 F}{\partial u_i \partial u_j}\, \delta u_i\, \delta u_j$$

Algebraic manipulation is used to separate the δy_i, $\delta y'_i$, and $\delta y''_i$ terms.

$$\begin{aligned}
\delta^2 F &= \frac{1}{2}\sum_{i=1}^{n}\sum_{j=1}^{3n}\frac{\partial^2 F}{\partial u_i \partial u_j}\delta u_i\,\delta u_j + \frac{1}{2}\sum_{i=n+1}^{2n}\sum_{j=1}^{3n}\frac{\partial^2 F}{\partial u_i \partial u_j}\delta u_i\,\delta u_j \\
&\quad + \frac{1}{2}\sum_{i=2n+1}^{3n}\sum_{j=1}^{3n}\frac{\partial^2 F}{\partial u_i \partial u_j}\delta u_i\,\delta u_j \\
&= \frac{1}{2}\sum_{i=1}^{n}\sum_{j=1}^{n}\frac{\partial^2 F}{\partial u_i \partial u_j}\delta u_i\,\delta u_j + \frac{1}{2}\sum_{i=1}^{n}\sum_{j=n+1}^{2n}\frac{\partial^2 F}{\partial u_i \partial u_j}\delta u_i\,\delta u_j \\
&\quad + \frac{1}{2}\sum_{i=1}^{n}\sum_{j=2n+1}^{3n}\frac{\partial^2 F}{\partial u_i \partial u_j}\delta u_i\,\delta u_j \\
&\quad + \frac{1}{2}\sum_{i=n+1}^{2n}\sum_{j=1}^{n}\frac{\partial^2 F}{\partial u_i \partial u_j}\delta u_i\,\delta u_j + \frac{1}{2}\sum_{i=n+1}^{2n}\sum_{j=n+1}^{2n}\frac{\partial^2 F}{\partial u_i \partial u_j}\delta u_i\,\delta u_j \\
&\quad + \frac{1}{2}\sum_{i=n+1}^{2n}\sum_{j=2n+1}^{3n}\frac{\partial^2 F}{\partial u_i \partial u_j}\delta u_i\,\delta u_j \\
&\quad + \frac{1}{2}\sum_{i=2n+1}^{3n}\sum_{j=1}^{n}\frac{\partial^2 F}{\partial u_i \partial u_j}\delta u_i\,\delta u_j + \frac{1}{2}\sum_{i=2n+1}^{3n}\sum_{j=n+1}^{2n}\frac{\partial^2 F}{\partial u_i \partial u_j}\delta u_i\,\delta u_j \\
&\quad + \frac{1}{2}\sum_{i=2n+1}^{3n}\sum_{j=2n+1}^{3n}\frac{\partial^2 F}{\partial u_i \partial u_j}\delta u_i\,\delta u_j
\end{aligned}$$

9.1. FIXED END POINTS

Changing back to the function variables results in

$$\begin{aligned}\delta^2 F &= \frac{1}{2}\sum_{i=1}^{n}\sum_{j=1}^{n}\frac{\partial^2 F}{\partial y_i \partial y_j}\delta y_i\, \delta y_j + \frac{1}{2}\sum_{i=1}^{n}\sum_{j=1}^{n}\frac{\partial^2 F}{\partial y_i \partial y'_j}\delta y_i\, \delta y'_j \\ &+ \frac{1}{2}\sum_{i=1}^{n}\sum_{j=1}^{n}\frac{\partial^2 F}{\partial y_i \partial y''_j}\delta y_i\, \delta y''_j \\ &+ \frac{1}{2}\sum_{i=1}^{n}\sum_{j=1}^{n}\frac{\partial^2 F}{\partial y'_i \partial y_j}\delta y'_i\, \delta y_j + \frac{1}{2}\sum_{i=1}^{n}\sum_{j=1}^{n}\frac{\partial^2 F}{\partial y'_i \partial y'_j}\delta y'_i\, \delta y'_j \\ &+ \frac{1}{2}\sum_{i=1}^{n}\sum_{j=1}^{n}\frac{\partial^2 F}{\partial y'_i \partial y''_j}\delta y'_i\, \delta y''_j \\ &+ \frac{1}{2}\sum_{i=1}^{n}\sum_{j=1}^{n}\frac{\partial^2 F}{\partial y''_i \partial y_j}\delta y''_i\, \delta y_j + \frac{1}{2}\sum_{i=1}^{n}\sum_{j=1}^{n}\frac{\partial^2 F}{\partial y''_i \partial y'_j}\delta y''_i\, \delta y'_j \\ &+ \frac{1}{2}\sum_{i=1}^{n}\sum_{j=1}^{n}\frac{\partial^2 F}{\partial y''_i \partial y''_j}\delta y''_i\, \delta y''_j\end{aligned}$$

Using the symmetry of the Hessian matrix gives

$$\begin{aligned}\delta^2 F &= \frac{1}{2}\sum_{i=1}^{n}\sum_{j=1}^{n}\frac{\partial^2 F}{\partial y_i \partial y_j}\delta y_i\, \delta y_j + \frac{1}{2}\sum_{i=1}^{n}\sum_{j=1}^{n}\frac{\partial^2 F}{\partial y'_i \partial y'_j}\delta y'_i\, \delta y'_j \\ &+ \frac{1}{2}\sum_{i=1}^{n}\sum_{j=1}^{n}\frac{\partial^2 F}{\partial y''_i \partial y''_j}\delta y''_i\, \delta y''_j \\ &+ \sum_{i=1}^{n}\sum_{j=1}^{n}\frac{\partial^2 F}{\partial y_i \partial y'_j}\delta y_i\, \delta y'_j + \sum_{i=1}^{n}\sum_{j=1}^{n}\frac{\partial^2 F}{\partial y'_i \partial y''_j}\delta y'_i\, \delta y''_j \\ &+ \sum_{i=1}^{n}\sum_{j=1}^{n}\frac{\partial^2 F}{\partial y_i \partial y''_j}\delta y_i\, \delta y''_j\end{aligned}$$

Inserting the expression for $\delta^2 F$ into $\delta^2 J$ gives

$$\begin{aligned}\delta^2 J &= \frac{1}{2}\int_a^b \sum_{i=1}^n \sum_{j=1}^n \left[\frac{\partial^2 F}{\partial y_i^2}\delta y_i\, \delta y_j + \frac{\partial^2 F}{\partial (y_i')}\delta y_i'\, \delta y_j' + \frac{\partial^2 F}{\partial (y_i'')^2}\delta y_i''\, \delta y_j''\right] dx \\ &+ \int_a^b \sum_{i=1}^n \sum_{j=1}^n \left[\frac{\partial^2 F}{\partial y_i \partial y_j'}\delta y_i\, \delta y_j' + \frac{\partial^2 F}{\partial y_i' \partial y_j''}\delta y_i'\, \delta y_j'' + \frac{\partial^2 F}{\partial y_i \partial y_j''}\delta y_i\, \delta y_j''\right] dx\end{aligned}$$
(9.1.6)

9.2 Variable End Points

We now consider the more general functional

$$J[\mathbf{y}] = \int_a^b F(x, \mathbf{y}, \mathbf{y}', \mathbf{y}'')\, dx + f(\mathbf{y}(b)) - g(\mathbf{y}(a)) + h(\mathbf{y}'(b)) - p(\mathbf{y}'(a)) \quad (9.2.1)$$

where F has continuous partial derivatives up to order 3 in each of its arguments. f, g, h, and p map $\mathbb{R}^n$ to $\mathbb{R}$ and $f, g, h, p \in \mathcal{C}^1(\mathbb{R}^n)$. The values of the functions at the endpoints are no longer fixed. The corresponding set of possible solutions $\mathcal{U}$ is

$$\mathcal{U} = \left\{\mathbf{y} : I \to \mathbb{R} \,\middle|\, y_i \in \mathcal{C}^2(I),\quad i = 1, \ldots, n\right\}$$

The minimization problem is

$$\mathbf{y}^* = \arg\operatorname*{local\,min}_{\mathbf{y} \in \mathcal{U}} J[\mathbf{y}] \quad (9.2.2)$$

The variations at the endpoints are arbitrary and hence $\mathcal{V} = \mathcal{U}$.

Extending the derivation of the previous section produces

$$\begin{aligned}
\delta J &= \int_a^b \delta F\, dx + \delta f - \delta g + \delta h - \delta p \\
&= \sum_{i=1}^n \int_a^b \left[\frac{\partial F}{\partial y_i}\delta y_i + \frac{\partial F}{\partial y_i'}\delta y_i' + \frac{\partial F}{\partial y_i''}\delta y_i''\right] dx \\
&\quad + \sum_{i=1}^n \left[\frac{\partial f}{\partial y_i}\delta y_i\right]\bigg|_b - \sum_{i=1}^n \left[\frac{\partial g}{\partial y_i}\delta y_i\right]\bigg|_a \\
&\quad + \sum_{i=1}^n \left[\frac{\partial h}{\partial y_i'}\delta y_i'\right]\bigg|_b - \sum_{i=1}^n \left[\frac{\partial p}{\partial y_i'}\delta y_i'\right]\bigg|_a \\
&= \sum_{i=1}^n \int_a^b \left[\frac{\partial F}{\partial y_i} - \frac{d}{dx}\frac{\partial F}{\partial y_i'} + \frac{d^2}{dx^2}\frac{\partial F}{\partial y_i''}\right]\delta y_i\, dx \\
&\quad + \sum_{i=1}^n \left[\frac{\partial F}{\partial y_i'}\delta y_i\right]\bigg|_a^b - \sum_{i=1}^n \left[\frac{d}{dx}\frac{\partial F}{\partial y_i''}\delta y_i\right]\bigg|_a^b \\
&\quad + \sum_{i=1}^n \left[\frac{\partial f}{\partial y_i}\delta y_i\right]\bigg|_b - \sum_{i=1}^n \left[\frac{\partial g}{\partial y_i}\delta y_i\right]\bigg|_a \\
&\quad + \sum_{i=1}^n \left[\frac{\partial F}{\partial y_i''}\delta y_i'\right]\bigg|_a^b + \sum_{i=1}^n \left[\frac{\partial h}{\partial y_i'}\delta y_i'\right]\bigg|_b - \sum_{i=1}^n \left[\frac{\partial p}{\partial y_i'}\delta y_i'\right]\bigg|_a
\end{aligned}$$

$$\delta J = \delta J_1 + \phi(b) - \psi(a) + \gamma(b) - \mu(a) \qquad (9.2.3)$$

$$\delta J_1 = \sum_{i=1}^{n} \int_a^b \left[\frac{\partial F}{\partial y_i} - \frac{d}{dx}\frac{\partial F}{\partial y'_i} + \frac{d^2}{dx^2}\frac{\partial F}{\partial y''_i} \right] \delta y_i \, dx$$

$$\phi(b) = \sum_{i=1}^{n} \left[\left(\frac{\partial F}{\partial y'_i} - \frac{d}{dx}\frac{\partial F}{\partial y''_i} + \frac{\partial f}{\partial y_i} \right) \delta y_i \right]\bigg|_{x=b}$$

$$\psi(a) = \sum_{i=1}^{n} \left[\left(\frac{\partial F}{\partial y'_i} - \frac{d}{dx}\frac{\partial F}{\partial y''_i} + \frac{\partial g}{\partial y_i} \right) \delta y_i \right]\bigg|_{x=a}$$

$$\gamma(b) = \sum_{i=1}^{n} \left[\left(\frac{\partial F}{\partial y''_i} + \frac{\partial h}{\partial y'_i} \right) \delta y'_i \right]\bigg|_{x=b}, \quad \mu(a) = \sum_{i=1}^{n} \left[\left(\frac{\partial F}{\partial y''_i} + \frac{\partial p}{\partial y'_i} \right) \delta y'_i \right]\bigg|_{x=a}$$

If the variations vanish at the endpoints, then $\phi(b) = \psi(a) = \gamma(b) = \mu(a) = 0$. The condition $\delta J = 0$ implies $\delta J_1 = 0$ and we recover the differential equations (9.1.5). Thus, the term in brackets for δJ_1 must vanish. If the variations at the endpoints do not vanish, then we must have

$$\left[\frac{\partial F}{\partial y'_i} - \frac{d}{dx}\frac{\partial F}{\partial y''_i} + \frac{\partial f}{\partial y_i} \right]\bigg|_b = 0, \quad \left[\frac{\partial F}{\partial y'_i} - \frac{d}{dx}\frac{\partial F}{\partial y''_i} + \frac{\partial g}{\partial y_i} \right]\bigg|_a = 0, \quad i = 1, \ldots, n$$

$$\left[\frac{\partial F}{\partial y''_i} + \frac{\partial h}{\partial y'_i} \right]\bigg|_b = 0, \quad \left[\frac{\partial F}{\partial y''_i} + \frac{\partial p}{\partial y'_i} \right]\bigg|_a = 0, \quad i = 1, \ldots, n$$

We have proved the following theorem.

Theorem 9.2.1. *Consider the closed interval $I = [a, b]$, the set of functions*

$$\mathcal{U} = \{ \mathbf{y} : I \to \mathbb{R} \mid y_i \in \mathcal{C}^2(I), \quad i = 1, \ldots, n \}$$

and the functional $J : \mathcal{U} \to \mathbb{R}$ defined by

$$J[\mathbf{y}] = \int_a^b F(x, \mathbf{y}, \mathbf{y}', \mathbf{y}'') \, dx + f(\mathbf{y}(b)) - g(\mathbf{y}(a)) + h(\mathbf{y}'(b)) - p(\mathbf{y}'(a)) \qquad (9.2.4)$$

where F has continuous partial derivatives up to order 3 and $f, g, h, p \in \mathcal{C}^1(\mathbb{R}^n)$. A necessary condition for a local minimum of J over $\mathcal{U}$ is that the following equations be satisfied:

$$\frac{\partial F}{\partial y_i} - \frac{d}{dx}\frac{\partial F}{\partial y'_i} + \frac{d^2}{dx^2}\frac{\partial F}{\partial y''_i} = 0, \quad i = 1, \ldots, n \qquad (9.2.5)$$

$$\left[\frac{\partial F}{\partial y'_i} - \frac{d}{dx}\frac{\partial F}{\partial y''_i} + \frac{\partial f}{\partial y_i} \right]\bigg|_b = 0, \quad \left[\frac{\partial F}{\partial y'_i} - \frac{d}{dx}\frac{\partial F}{\partial y''_i} + \frac{\partial g}{\partial y_i} \right]\bigg|_a = 0, \quad i = 1, \ldots, n \qquad (9.2.6)$$

$$\left[\frac{\partial F}{\partial y''_i} + \frac{\partial h}{\partial y'_i} \right]\bigg|_b = 0, \quad \left[\frac{\partial F}{\partial y''_i} + \frac{\partial p}{\partial y'_i} \right]\bigg|_a = 0, \quad i = 1, \ldots, n \qquad (9.2.7)$$

9.2. VARIABLE END POINTS

The 2nd variation $\delta^2 J$ is easily computed from the results of the last section. Assume that $f, g, h, p \in C^2(\mathbb{R}^n)$.

$$\begin{aligned}
\delta^2 J &= \frac{1}{2} \int_a^b \sum_{i=1}^n \sum_{j=1}^n \left[\frac{\partial^2 F}{\partial y_i^2} \delta y_i\, \delta y_j + \frac{\partial^2 F}{\partial (y_i')} \delta y_i'\, \delta y_j' + \frac{\partial^2 F}{\partial (y_i'')^2} \delta y_i''\, \delta y_j'' \right] dx \\
&+ \int_a^b \sum_{i=1}^n \sum_{j=1}^n \left[\frac{\partial^2 F}{\partial y_i \partial y_j'} \delta y_i\, \delta y_j' + \frac{\partial^2 F}{\partial y_i' \partial y_j''} \delta y_i'\, \delta y_j'' + \frac{\partial^2 F}{\partial y_i \partial y_j''} \delta y_i\, \delta y_j'' \right] dx \\
&+ \frac{1}{2} \sum_{i=1}^n \sum_{j=1}^n \left[\frac{\partial^2 f}{\partial y_i \partial y_j} \delta y_i\, \delta y_j \right]\bigg|_b - \frac{1}{2} \sum_{i=1}^n \sum_{j=1}^n \left[\frac{\partial^2 g}{\partial y_i \partial y_j} \delta y_i\, \delta y_j \right]\bigg|_a \\
&+ \frac{1}{2} \sum_{i=1}^n \sum_{j=1}^n \left[\frac{\partial^2 h}{\partial y_i' \partial y_j'} \delta y_i'\, \delta y_j' \right]\bigg|_b - \frac{1}{2} \sum_{i=1}^n \sum_{j=1}^n \left[\frac{\partial^2 p}{\partial y_i' \partial y_j'} \delta y_i'\, \delta y_j' \right]\bigg|_a \quad (9.2.8)
\end{aligned}$$

Part IV
Classical Mechanics

Chapter 10

Introduction

10.1 Newtonian Dynamics

The remainder of this book deals with applications of calculus of variations to classical mechanics. We begin by providing a very brief summary of Newtonian dynamics. Extensive information on this subject is found in many texts such as [25, 30].

For the sake of simplicity, we consider objects as point particles. In other words, internal structure and spatial orientation are not required to describe the dynamics of the object. More complete accounts of mechanics are provided in other texts (see [10, 25, 30] for details).

The center of mass is usually chosen to be the position of the object. Let $\mathbf{r}(t)$ be the position vector of an object in a 3-dimensional cartesian coordinate system. The velocity $\mathbf{v}(t)$ and acceleration $\mathbf{a}(t)$ of the object are given by

$$\mathbf{v}(t) = \dot{\mathbf{r}}(t), \qquad \mathbf{a}(t) = \dot{\mathbf{v}}(t) = \ddot{\mathbf{r}}(t) \qquad (10.1.1)$$

where the dots indicate time derivatives.

The Newtonian formulation of classical mechanics is based on Newton's laws of motion.

- 1st law: An object moves with a constant velocity unless acted upon by a net force.

- 2nd law: The net force acting on an object equals the mass of the object times its acceleration.

- 3rd law: Action equals reaction.

Newton's laws are formulated in an inertial frame. The 2nd law is stated in vector form as: $\mathbf{F} = m\mathbf{a}$. This law governs the dynamics of the object. The word *dynamics* means that we wish to compute the future position and velocity of an object. Knowledge of the force allows us to compute acceleration. The future position and velocity are then determined by integration.

$$\mathbf{a}(t) = \frac{1}{m}\mathbf{F}(t) \tag{10.1.2}$$

$$\mathbf{v}(t) = \int_{t_0}^{t} \mathbf{a}(\tau)\,d\tau + \mathbf{v}_0 \tag{10.1.3}$$

$$\mathbf{r}(t) = \int_{t_0}^{t} \mathbf{v}(\tau)\,d\tau + \mathbf{r}_0 \tag{10.1.4}$$

where t_0 is the initial time and t is the current time. The initial position $\mathbf{r}(t_0) = \mathbf{r}_0$ and initial velocity $\mathbf{v}(t_0) = \mathbf{v}_0$ are constants of integration which uniquely determine the solution.

$\mathbf{F}$ is called a *conservative force* if it can be derived from a potential energy function U in the following way:

$$\mathbf{F}(\mathbf{r}) = -\nabla U(\mathbf{r}) \tag{10.1.5}$$

An object in a conservative force field experiences force $\mathbf{F}(\mathbf{r})$ when $\mathbf{r}(t) = \mathbf{r}$.

A *conservative system* obeys the *conservation of total energy*. For a single object, this is expressed as

$$E = T(\dot{\mathbf{r}}) + U(\mathbf{r}) = \text{constant} \tag{10.1.6}$$

where $T(\dot{\mathbf{r}}) = \frac{1}{2}m\|\mathbf{v}\|_2^2$ is the kinetic energy of the object.

For a system of n objects, the equations of Newtonian dynamics are

$$\mathbf{a}_i(t) = \frac{1}{m_i}\mathbf{F}_i(t) \tag{10.1.7}$$

$$\mathbf{v}_i(t) = \int_{t_0}^{t} \mathbf{a}_i(\tau)\,d\tau + \mathbf{v}_{i0} \tag{10.1.8}$$

$$\mathbf{r}_i(t) = \int_{t_0}^{t} \mathbf{v}_i(\tau)\,d\tau + \mathbf{r}_{i0} \tag{10.1.9}$$

The index $i = 1, \ldots, n$ indicates the specific object. The net force $\mathbf{F}_i(t)$ is the force acting on object i. This force can be partitioned into pairwise forces $\mathbf{F}_{ij}$ between objects and an external force $\mathbf{F}_{\text{ext}}^{(i)}$.

$$\mathbf{F}_i(t) = \sum_{j \neq i}^{n} \mathbf{F}_{ij}(t) + \mathbf{F}_{\text{ext}}^{(i)}(t) \tag{10.1.10}$$

Newton's 3rd law asserts that $\mathbf{F}_{ij} = -\mathbf{F}_{ji}$.

10.1. NEWTONIAN DYNAMICS

The total energy E of a conservative system is

$$E = \frac{1}{2} \sum_{i=1}^{n} m_i \|\mathbf{v}_i\|_2^2 + U(\mathbf{r}_1, \ldots, \mathbf{r}_n) = \text{constant} \tag{10.1.11}$$

where U is the potential energy of the combined system.

The center of mass $\mathbf{R}$ of the system is defined by

$$\mathbf{R} = \frac{1}{M} \sum_{i=1}^{n} m_i \mathbf{r}_i, \qquad M = \sum_{i=1}^{n} m_i \tag{10.1.12}$$

where M is the total mass. The velocity $\mathbf{V}$ and acceleration $\mathbf{A}$ of the center of mass are computed by the obvious definitions

$$\mathbf{V} = \dot{\mathbf{R}}, \qquad \mathbf{A} = \dot{\mathbf{V}} = \ddot{\mathbf{R}} \tag{10.1.13}$$

The motion of the center of mass is determined by the equation

$$M\mathbf{A} = \sum_{i=1}^{n} \mathbf{F}_{\text{ext}}^{(i)} \tag{10.1.14}$$

This is Newton's 2nd law applied to the center of mass. Notice that only external forces are involved in the sum. The internal forces, being pairwise equal and opposite, completely cancel.

The momentum of an object is defined by $\mathbf{p}_i = m_i \mathbf{v}_i$. The total momentum of the system is

$$\mathbf{P} = \sum_{i=1}^{n} \mathbf{p}_i = \sum_{i=1}^{n} m_i \mathbf{v}_i = M\mathbf{V} \tag{10.1.15}$$

When the net external force on the system is 0, then according to Newton's 2nd law: $M\mathbf{A} = \dot{\mathbf{P}} = \mathbf{0}$. In this case, we have the *conservation of linear momentum*.

$$\mathbf{P} = \sum_{i=1}^{n} \mathbf{p}_i = \sum_{i=1}^{n} m_i \mathbf{v}_i = \text{constant} \tag{10.1.16}$$

The total angular momentum **L** of the system is

$$\mathbf{L} = \sum_{i=1}^{n} (\mathbf{r}_i \times \mathbf{p}_i) \tag{10.1.17}$$

The total external torque τ_{ext} is defined as

$$\tau_{\text{ext}} = \sum_{i=1}^{n} \left(\mathbf{r}_i \times \mathbf{F}_{\text{ext}}^{(i)} \right) \tag{10.1.18}$$

The relationship between torque and angular momentum is

$$\tau_{\text{ext}} = \dot{\mathbf{L}} \tag{10.1.19}$$

When there is no external torque applied to the system, then $\dot{\mathbf{L}} = \mathbf{0}$ and *conservation of angular momentum* holds.

$$\mathbf{L} = \sum_{i=1}^{n} (\mathbf{r}_i \times \mathbf{p}_i) = \text{constant} \tag{10.1.20}$$

One major problem with Newtonian mechanics is the difficulty dealing with non-cartesian coordinate systems. The equations become more complex and cumbersome to solve. Another issue is that the dynamics of a system of objects requires specification of the forces between each pair of objects. In many situations, it is difficult to compute forces but easier to specify energy. The goal of the remainder of this book is to introduce alternative formulations of classical mechanics.

$$T=\frac{1}{2}m(\dot{x})^2 \quad U=\frac{1}{2}k x^2 \qquad T=\frac{1}{2}m l^2(\dot{\theta})^2 \quad U=mgl(1-\cos\theta)$$

Mass-Spring   **Pendulum**

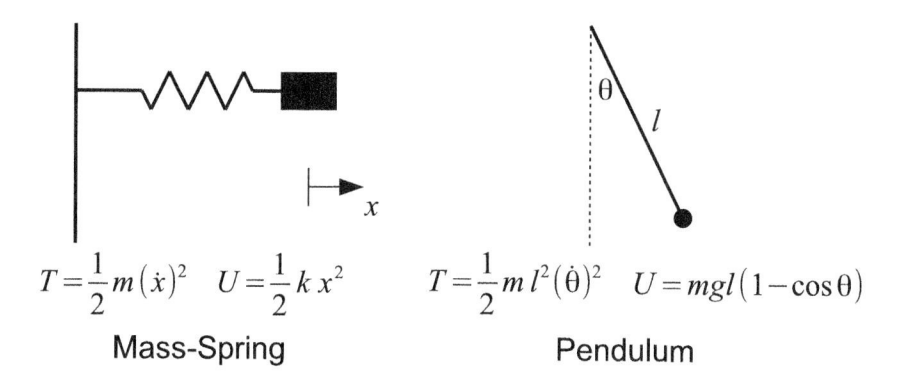

Figure 10.1.1: Examples of generalized coordinates

10.2 Generalized Coordinates

The Lagrangian and Hamiltonian formulation of classical mechanics are alternatives to the usual Newtonian formulation. They provide identical solutions to Newtonian mechanics for the same problem. However, there are some advantages of these alternative methods. The primary advantage is they can easily handle generalized coordinates. Generalized coordinates are any coordinate system which sufficiently describes the system.

For the mass-spring system in figure 10.1.1, the cartesian coordinate x and the velocity $\dot{x}$ adequately determine the systems current configuration. The plane pendulum in figure 10.1.1 is described by the angle θ between the pendulum and the vertical axis. Notice that θ is not a cartesian coordinate but is a generalized coordinate which describes the system. The corresponding generalized velocity for this system is $\dot{\theta}$.

Chapter 11

Single DOF Systems

11.1 Lagrangian Dynamics

A single DOF (Degree Of Freedom) system is specified by a single generalized coordinate $q(t)$. The generalized coordinate $q(t)$ and its corresponding generalized velocity $\dot{q}(t)$ completely determine the current state of the system. We begin by defining a function which contains the kinetic and potential energies.

Definition 11.1.1 (Lagrangian). The *Lagrangian* L of a single DOF system is defined as
$$L(q,\dot{q},t) = T(q,\dot{q},t) - U(q,\dot{q},t) \tag{11.1.1}$$

T is the kinetic energy and U is the potential energy. Notice that the Lagrangian L is the difference of these energies and not the sum. Hence, L is not the total energy of the system. This may seem odd at first, but we shall be able to make a variational statement concerning L.

In this chapter, we will assume the kinetic energy has the form
$$T(q,\dot{q}) = \frac{1}{2}a(q)\,\dot{q}^2 \tag{11.1.2}$$

where $a(q) > 0$. The function $a(q)$ is linearly proportional to the mass m of the system. It will also be assumed that the potential energy $U(q)$ is a function of position only. With these assumptions, the Lagrangian $L(q,\dot{q})$ is not an explicit function of time t and has the form
$$L(q,\dot{q}) = \frac{1}{2}a(q)\,\dot{q}^2 - U(q) \tag{11.1.3}$$

Lagrangian dynamics is based on a variational statement known as *Hamilton's principle*. Let t_0 be the initial time and t_f be the final time and suppose that $q(t_0)$ and $q(t_f)$ are known. We assume that $q \in \mathcal{C}^1([t_0, t_f])$.

Theorem 11.1.1 (Hamilton's Principle). *The dynamical path $q(t)$ of a classical single DOF system from $q(t_0)$ to $q(t_f)$ is a stationary point of the functional*

$$S[q] = \int_{t_0}^{t_f} L(q, \dot{q}, t)\, dt \qquad (11.1.4)$$

In other words, $\delta S = 0$.

The functional S is called the *action*. Hamilton's principle is proved in [10, 21] using D'Alembert's principle. The reader should refer to those texts for details. Notice that only a stationary point of the functional is required and not a minimum. For certain Lagrangians, it can be shown that the functional achieves a minimum. The Euler differential equation (5.1.5) associated with the functional (11.1.4) is

$$\frac{\partial L}{\partial q} - \frac{d}{dt}\frac{\partial L}{\partial \dot{q}} = 0$$

Multiplying through by -1 gives the conventional form of Lagrange's equation of motion.

Theorem 11.1.2 (Lagrange's Equation). *The dynamical path $q(t)$ of a classical single DOF system from $q(t_0)$ to $q(t_f)$ is a solution to the differential equation*

$$\frac{d}{dt}\frac{\partial L}{\partial \dot{q}} - \frac{\partial L}{\partial q} = 0 \qquad (11.1.5)$$

We now discuss a subtle point concerning Lagrange's equation. Knowledge of the final position $q(t_f)$ was assumed by Hamilton's principle which in turn allowed the derivation of Lagrange's equation. However, equation (11.1.5) is a 2nd order ordinary differential equation which has a unique solution $q(t)$ when provided with an initial position $q(t_0)$ and initial velocity $\dot{q}(t_0)$. Computing such a solution then gives the final position $q(t_f)$. Having both the initial and final positions allows us to apply Hamilton's principle and thus confirms that the solution is indeed correct.

Lagrange's equation for the system (11.1.3) is

$$a(q)\ddot{q} + \frac{1}{2}a'(q)\dot{q}^2 + U'(q) = 0 \qquad (11.1.6)$$

This system satisfies the Legendre condition (5.1.12).

$$\frac{\partial^2 L}{\partial \dot{q}^2} = a(q) > 0$$

11.2 Conservation of Energy

Assume that the Lagrangian $L(q, \dot{q})$ is not an explicit function of time. We compute the total time derivative of the Lagrangian and use Lagrange's equation (11.1.5) to achieve the following result.

$$\frac{dL}{dt} = \frac{\partial L}{\partial q}\dot{q} + \frac{\partial L}{\partial \dot{q}}\ddot{q}$$

$$\frac{dL}{dt} = \left(\frac{d}{dt}\frac{\partial L}{\partial \dot{q}}\right)\dot{q} + \frac{\partial L}{\partial \dot{q}}\ddot{q}$$

$$\frac{dL}{dt} = \frac{d}{dt}\left[\frac{\partial L}{\partial \dot{q}}\dot{q}\right]$$

$$0 = \frac{d}{dt}\left[\frac{\partial L}{\partial \dot{q}}\dot{q}\right] - \frac{dL}{dt}$$

$$0 = \frac{d}{dt}\left[\frac{\partial L}{\partial \dot{q}}\dot{q} - L\right]$$

The term in brackets is always constant in time. Such quantities are called *constants of motion* or *integrals of motion*. For a Lagrangian which is not an explicit function of time, the term in brackets is the total energy E of the system.

$$E = \frac{\partial L}{\partial \dot{q}}\dot{q} - L = \text{constant} \tag{11.2.1}$$

This equation is called the *conservation of energy*. A system which conserves energy is called *conservative*.

We can explicitly demonstrate the conservation of energy for the Lagrangian given by equation (11.1.3). Computation of the partial derivative of L with respect to $\dot{q}$ gives

$$\frac{\partial L}{\partial \dot{q}} = a(q)\,\dot{q}$$

Using this value, we now have

$$\frac{d}{dt}\left[\frac{\partial L}{\partial \dot{q}}\dot{q} - L\right] = 0$$

$$\frac{d}{dt}\left[a(q)\,\dot{q}^2 - \left[\frac{1}{2}a(q)\,\dot{q}^2 - U(q)\right]\right] = 0$$

$$\frac{d}{dt}\left[\frac{1}{2}a(q)\,\dot{q}^2 + U(q)\right] = 0$$

$$\frac{d}{dt}[T(q,\dot{q}) + U(q)] = 0$$

This confirms the conservation of total energy.

$$E = \frac{\partial L}{\partial \dot{q}} \dot{q} - L = T + U = \text{constant}$$

11.3 Momentum

The generalized momentum p of a single DOF system is defined as

$$p = \frac{\partial L}{\partial \dot{q}} \qquad (11.3.1)$$

For equation (11.1.3), the momentum is computed to be

$$p = a(q)\,\dot{q} \qquad (11.3.2)$$

Using Lagrange's equation (11.1.5), we observe that

$$\frac{\partial L}{\partial q} = \frac{d}{dt}\frac{\partial L}{\partial \dot{q}} = \dot{p}$$

The generalized momentum p is a constant of motion when $\dot{p} = 0$. This is seen in terms of the Lagrangian as

$$\dot{p} = \frac{\partial L}{\partial q} = 0 \qquad (11.3.3)$$

If the Lagrangian is not an explicit function of q, then we have the *conservation of momentum*.

We can define an associated generalized force F using

$$F = \frac{\partial L}{\partial q} = \dot{p} \qquad (11.3.4)$$

Equation (11.3.4) is a generalized version of Newton's 2nd law: $F = ma$. The analogy is exact when $q = x$ is a cartesian coordinate, the force F is conservative, and linear momentum is $p = mv$. The Lagrangian for this case is

$$L = \frac{1}{2}mv^2 - U(x)$$

We compute

$$F = \frac{\partial L}{\partial x} = -U'(x), \qquad p = \frac{\partial L}{\partial v} = mv$$

We easily see that: $F = \dot{p} = m\dot{v} = ma$. When $F = 0$, then $\dot{p} = 0$ and the momentum $p = mv$ is conserved.

11.4 Hamiltonian Dynamics

Hamiltonian dynamics provides another alternative to Newtonian mechanics. Hamilton's equations are derived in this section for the single DOF system using the *Legendre transformation*. Consider the total differential of the Lagrangian.

$$dL = \frac{\partial L}{\partial q}dq + \frac{\partial L}{\partial \dot{q}}d\dot{q} + \frac{\partial L}{\partial t}dt \tag{11.4.1}$$

Using equations (11.3.1) and (11.3.4), we have

$$dL = \dot{p}\,dq + p\,d\dot{q} + \frac{\partial L}{\partial t}dt \tag{11.4.2}$$

The product rule of differentials is

$$d(p\dot{q}) = \dot{q}\,dp + p\,d\dot{q}$$
$$d(p\dot{q}) - \dot{q}\,dp = p\,d\dot{q}$$

Using this result in equation (11.4.2) gives

$$dL = \dot{p}\,dq + d(p\dot{q}) - \dot{q}\,dp + \frac{\partial L}{\partial t}dt$$
$$-dL = -\dot{p}\,dq - d(p\dot{q}) + \dot{q}\,dp - \frac{\partial L}{\partial t}dt$$
$$d(p\dot{q}) - dL = -\dot{p}\,dq + \dot{q}\,dp - \frac{\partial L}{\partial t}dt$$
$$d[p\dot{q} - L] = -\dot{p}\,dq + \dot{q}\,dp - \frac{\partial L}{\partial t}dt$$

Define $H = p\dot{q} - L$. H is called the Hamiltonian of the system and it is assumed to be a function of q and p. The right hand side, of the expressions above, is a perfect differential if we write the following set of equations and match differentials:

$$d[p\dot{q} - L] = -\dot{p}\,dq + \dot{q}\,dp - \frac{\partial L}{\partial t}dt$$
$$dH = \frac{\partial H}{\partial q}dq + \frac{\partial H}{\partial p}dp + \frac{\partial H}{\partial t}dt$$

Thus, we have

$$\frac{\partial H}{\partial q} = -\dot{p}, \qquad \frac{\partial H}{\partial p} = \dot{q}, \qquad \frac{\partial H}{\partial t} = -\frac{\partial L}{\partial t}$$

These equations are called Hamilton's equations of motion.

Definition 11.4.1 (Hamiltonian). The *Hamiltonian* $H(q,p,t)$ of a single DOF system is defined in terms of the Lagrangian as

$$H = p\dot{q} - L \tag{11.4.3}$$

The Legendre transformation results in Hamilton's equation of motion.

Theorem 11.4.1 (Hamilton's Equations). *The dynamical path $q(t)$ and $p(t)$ of a classical single DOF system are solutions to the set of differential equations*

$$\dot{q} = \frac{\partial H}{\partial p} \tag{11.4.4}$$

$$\dot{p} = -\frac{\partial H}{\partial q} \tag{11.4.5}$$

Hamilton's equations are pair of 1st order ordinary differential equations. Two constants are integration are required to uniquely determine a solution. The initial position $q(t_0) = q_0$ and initial momentum $p(t_0) = p_0$ allow us to compute the constants of integration.

Thus far, we have only discussed a differential transformation from $L(q,\dot{q},t)$ to $H(q,p,t)$. We still need a way to convert from velocity $\dot{q}$ to momentum p. A procedure to accomplish this is demonstrated on the case of the Lagrangian of equation (11.1.3). First, compute the momentum p using equation (11.3.1) and solve for $\dot{q}$.

$$p = \frac{\partial L}{\partial \dot{q}} = a(q)\dot{q}, \qquad \dot{q} = \frac{p}{a(q)}$$

Using this result in equation (11.4.3) gives

$$\begin{aligned}
H &= p\dot{q} - \left[\frac{1}{2}a(q)\dot{q}^2 - U(q)\right] \\
&= \frac{p^2}{a(q)} - \left[\frac{1}{2}a(q)\left[\frac{p}{a(q)}\right]^2 - U(q)\right] \\
&= \frac{p^2}{a(q)} - \left[\frac{p^2}{2a(q)} - U(q)\right] \\
&= \frac{p^2}{2a(q)} + U(q)
\end{aligned}$$

The first term is observed to be the kinetic energy in terms of p.

$$T(p,q) = \frac{p^2}{2a(q)}$$

This system was shown to be conservative in a previous section. The Hamiltonian is the total energy of the system.

$$H(q,p) = T(q,p) + U(q) = E = \text{constant} \tag{11.4.6}$$

11.4. HAMILTONIAN DYNAMICS

Hamilton's equation of motion for the system are

$$\dot{q} = \frac{p}{a(q)}$$
$$\dot{p} = \frac{p^2}{2} \frac{a'(q)}{[a(q)]^2} - U'(q)$$

When $q = x$ as the usual 1-dimensional cartesian coordinate, we have

$$H = \frac{p^2}{2m} + U(x)$$

The equations of motion for this case are

$$\dot{x} = \frac{p}{m}$$
$$\dot{p} = -U'(x)$$

11.5 Integration of the Equation of Motion

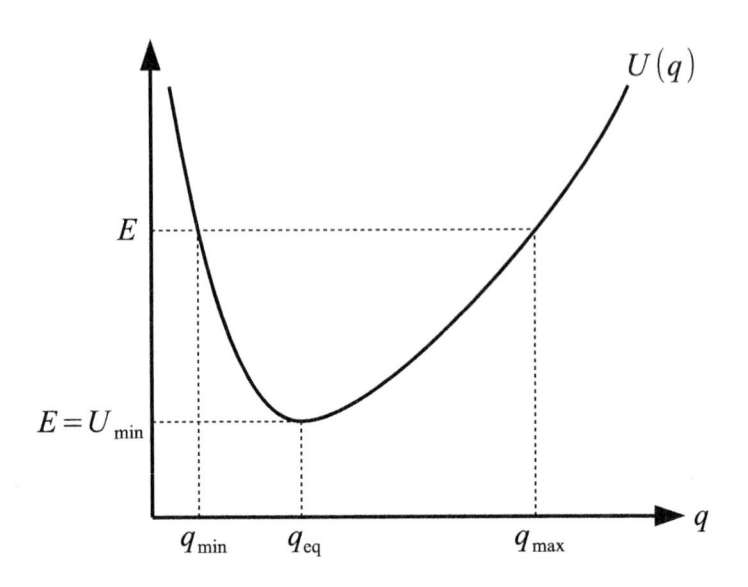

Figure 11.5.1: $U(q)$ for a bounded system

Consider a bounded conservative system where $q_{min} \leq q(t) \leq q_{max}$ for all $t \geq t_0$ and $U(q)$ has the shape depicted in figure 11.5.1. We can bypass solving any equations of motion. Instead, we obtain the solution directly from the conservation of energy.

$$\begin{aligned} T(q,\dot{q}) + U(q) &= E \\ T(q,\dot{q}) &= E - U(q) \\ \frac{1}{2} a(q) \dot{q}^2 &= E - U(q) \\ \dot{q}^2 &= \frac{2}{a(q)} [E - U(q)] \end{aligned}$$

For a bounded system, we must have $E \geq U(q)$ and can take the square root of both sides of the equation above.

$$\dot{q} = \text{sign}(\dot{q}) \sqrt{\frac{2}{a(q)} [E - U(q)]}$$

11.5. INTEGRATION OF THE EQUATION OF MOTION

Suppose $t_0 \leq t \leq t_1$ such that $\dot{q}(t)$ does not change sign from t_0 to t_1. We separate by variables and integrate.

$$\frac{dq}{dt} = \text{sign}(\dot{q})\sqrt{\frac{2}{a(q)}[E - U(q)]}$$

$$dt = \frac{\text{sign}(\dot{q})}{\sqrt{2}}\sqrt{\frac{a(q)}{E - U(q)}}\, dq$$

$$\int_{t_0}^{t} d\tau = \frac{\text{sign}(\dot{q})}{\sqrt{2}} \int_{q(t_0)}^{q(t)} \sqrt{\frac{a(q)}{E - U(q)}}\, dq$$

The solution $q(t)$ is given implicitly by

$$t - t_0 = \frac{\text{sign}(\dot{q})}{\sqrt{2}} \int_{q(t_0)}^{q(t)} \sqrt{\frac{a(q)}{E - U(q)}}\, dq \qquad (11.5.1)$$

Any appropriate quadrature technique can be used to compute $q(t)$.

Consider the special case when $E = U_{\min} = U(q_{\text{eq}})$ as shown in figure 11.5.1. In this case we have $\dot{q}(t) = 0$ for all time t and the system remains at q_{eq}. This is known as an *equilibrium point*. The stability of equilibrium points is discussed in texts on differential equations (see [1, 15, 26] for details). We simply note the following results without proof:

$U''(q_{\text{eq}}) > 0 \;\Rightarrow\;$ Stable equilibrium
$U''(q_{\text{eq}}) = 0 \;\Rightarrow\;$ Further investigation required (see discussion below)
$U''(q_{\text{eq}}) < 0 \;\Rightarrow\;$ Unstable equilibrium

Some conclusions can be drawn when $U''(q_{\text{eq}}) = 0$.
If $U^{(k)}(q_{\text{eq}}) = 0$ for all $k \geq 2$, then we have neutral equilibrium.
Suppose $n > 2$ is the lowest order derivative which is non–zero at q_{eq}.

$$U^{(n)}(q_{\text{eq}}) \neq 0 \quad \text{and} \quad U^{(k)}(q_{\text{eq}}) = 0, \quad k = 1, \ldots, n-1$$

If n is even, then we have

$U^{(n)}(q_{\text{eq}}) > 0 \;\Rightarrow\;$ Stable equilibrium
$U^{(n)}(q_{\text{eq}}) < 0 \;\Rightarrow\;$ Unstable equilibrium

No other definite conclusions can be drawn.

11.6 Mass-Spring System

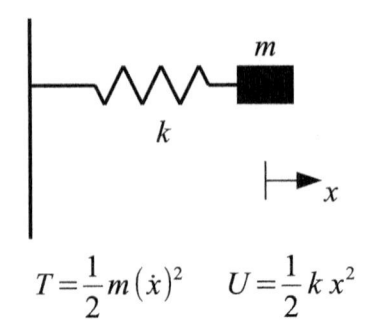

Figure 11.6.1: Mass-spring system

$$T = \frac{1}{2}m(\dot{x})^2 \quad U = \frac{1}{2}kx^2$$

For a comparison of Lagrange equations to Newton's 2nd law, we consider the example of a mass-spring system. The Lagrangian for this system is

$$L = \frac{1}{2}m\dot{x}^2 - \frac{1}{2}kx^2$$

where k is the spring constant. Using Lagrange's equation, we obtain

$$\frac{d}{dt}\frac{\partial L}{\partial \dot{x}} - \frac{\partial L}{\partial x} = 0$$
$$\frac{d}{dt}(m\dot{x}) - (-kx) = 0$$
$$m\ddot{x} + kx = 0$$

This is equivalent to Newton's 2nd law

$$F = m\ddot{x}, \quad F = -kx$$

The 2nd order differential equation is usually written in the form

$$\ddot{x} + \omega^2 x = 0, \quad \omega = \sqrt{\frac{k}{m}}$$

The period of oscillation is: $T = 2\pi/\omega$. The solution is

$$x(t) = A\cos(\omega t + \phi), \quad A = \sqrt{x_0^2 + \left(\frac{\dot{x}_0}{\omega}\right)^2}, \quad \tan\phi = -\frac{\dot{x}_0}{\omega x_0}$$

where the amplitude A and phase ϕ are computed using the initial conditions $x(0) = x_0$ and $\dot{x}(0) = \dot{x}_0$.

11.7 Plane Pendulum

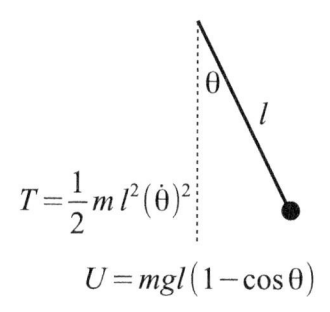

Figure 11.7.1: Plane pendulum

The plane pendulum problem, given in figure 11.7.1, can be solved using direct integration. Let θ_{max} be the maximum angle during the motion. The problem is symmetric and thus we can compute the period of oscillation T as 4 times the time it takes to go from $\theta = 0$ to $\theta = \theta_{max}$. The total energy is: $E = mgl[1 - \cos\theta_{max}]$. This value is obtained at the turning point of motion when $U = E$. We substitute the expressions for T, E, and U into equation (11.5.1).

$$T = \frac{4}{\sqrt{2}} \int_0^{\theta_{max}} \sqrt{\frac{ml^2}{mgl[\cos\theta - \cos\theta_{max}]}} \, d\theta$$

$$T = 4\sqrt{\frac{l}{2g}} \int_0^{\theta_{max}} \frac{d\theta}{\sqrt{\cos\theta - \cos\theta_{max}}}$$

This formula can be converted to an elliptic integral of the first kind. The details are found in [22, 25] and a series approximation is

$$T = 2\pi\sqrt{\frac{l}{g}} \left[1 + \frac{1}{16}\theta_{max}^2 + \frac{11}{3072}\theta_{max}^4 + \cdots \right]$$

For small angle oscillations, we get the familiar result

$$T = 2\pi\sqrt{\frac{l}{g}}$$

Chapter 12

Multiple DOF Systems

12.1 Lagrangian Dynamics

Very few problems in classical mechanics can accurately be described by a single DOF. Thus, we are compelled to discuss multiple DOF systems. Specific examples are provided later in this chapter. Let $q_1(t), \ldots, q_n(t)$ be the n generalized coordinates which describe the system. We will often use vector notation for these coordinates and their generalized velocities.

$$\mathbf{q} = [q_1, \ldots, q_n]^T, \qquad \dot{\mathbf{q}} = [\dot{q}_1, \ldots, \dot{q}_n]^T$$

With T and U representing kinetic and potential energy respectively, we define the Lagrangian for a multiple DOF system.

Definition 12.1.1 (Lagrangian). The *Lagrangian* L of a multiple DOF system is defined as

$$L(\mathbf{q}, \dot{\mathbf{q}}, t) = T(\mathbf{q}, \dot{\mathbf{q}}, t) - U(\mathbf{q}, \dot{\mathbf{q}}, t) \tag{12.1.1}$$

We will usually focus on a Lagrangian of a specific type. The kinetic energy is a quadratic form in velocity.

$$T(\mathbf{q}, \dot{\mathbf{q}}) = \frac{1}{2} \dot{\mathbf{q}}^T \mathbf{A}(\mathbf{q}) \, \dot{\mathbf{q}} = \frac{1}{2} \sum_{i=1}^{n} \sum_{j=1}^{n} A_{ij}(q_1, \ldots, q_n) \, \dot{q}_i \dot{q}_j \tag{12.1.2}$$

where $\mathbf{A}(\mathbf{q})$ is a symmetric positive definite matrix for every possible value of $\mathbf{q}$ for the system. Potential energy is a function of coordinates only: $U(\mathbf{q}) = U(q_1, \ldots, q_n)$. The Lagrangian for these energies is

$$L(\mathbf{q}, \dot{\mathbf{q}}) = \frac{1}{2} \dot{\mathbf{q}}^T \mathbf{A}(\mathbf{q}) \, \dot{\mathbf{q}} - U(\mathbf{q}) \tag{12.1.3}$$

Hamilton's principle for multiple DOF systems can be formulated if the initial point $\mathbf{q}(t_0) = \mathbf{q}_0$ and final point $\mathbf{q}(t_f) = \mathbf{q}_f$ are known. Assume that $q_i \in \mathcal{C}^1([t_0, t_f])$ for $i = 1, \ldots, n$.

Theorem 12.1.1 (Hamilton's Principle). *The dynamical path* $\mathbf{q}(t)$ *of a classical multiple DOF system from* $\mathbf{q}(t_0)$ *to* $\mathbf{q}(t_f)$ *is a stationary point of the functional*

$$S[\mathbf{q}] = \int_{t_0}^{t_f} L(\mathbf{q}, \dot{\mathbf{q}}, t)\, dt \tag{12.1.4}$$

In other words, $\delta S = 0$.

Equation (8.1.6) provides

$$\frac{\partial L}{\partial q_i} - \frac{d}{dt}\frac{\partial L}{\partial \dot{q}_i} = 0, \quad i = 1, \ldots, n$$

We reverse the signs to comply with our convention for Lagrange's equations.

Theorem 12.1.2 (Lagrange's Equations). *The dynamical path* $\mathbf{q}(t)$ *of a classical multiple DOF system from* $\mathbf{q}(t_0)$ *to* $\mathbf{q}(t_f)$ *is a solution to the differential equations*

$$\frac{d}{dt}\frac{\partial L}{\partial \dot{q}_i} - \frac{\partial L}{\partial q_i} = 0, \quad i = 1, \ldots, n \tag{12.1.5}$$

Lagrange's equations are a system of n 2nd order ordinary differential equations. A unique solution requires specifying $2n$ constants of integration and these are provided by the initial positions $\mathbf{q}_0$ and initial velocities $\dot{\mathbf{q}}_0$.

Lagrange's equations can also be written in vector form.

$$\frac{d}{dt}\frac{\partial L}{\partial \dot{\mathbf{q}}} - \frac{\partial L}{\partial \mathbf{q}} = \mathbf{0} \tag{12.1.6}$$

This page intentionally left blank.

12.2 Conservation of Energy

Consider the Lagrangian $L(\mathbf{q}, \dot{\mathbf{q}})$ which is not an explicit function of time. We compute the total time derivative of this Lagrangian and use Lagrange's equations (12.1.5).

$$\frac{dL}{dt} = \sum_{i=1}^{n} \frac{\partial L}{\partial q_i}\dot{q}_i + \sum_{i=1}^{n} \frac{\partial L}{\partial \dot{q}_i}\ddot{q}_i$$

$$\frac{dL}{dt} = \sum_{i=1}^{n} \left(\frac{d}{dt}\frac{\partial L}{\partial \dot{q}_i}\right)\dot{q}_i + \sum_{i=1}^{n} \frac{\partial L}{\partial \dot{q}_i}\ddot{q}_i$$

$$\frac{dL}{dt} = \frac{d}{dt}\left[\sum_{i=1}^{n} \frac{\partial L}{\partial \dot{q}_i}\dot{q}_i\right]$$

$$0 = \frac{d}{dt}\left[\sum_{i=1}^{n} \frac{\partial L}{\partial \dot{q}_i}\dot{q}_i\right] - \frac{dL}{dt}$$

$$0 = \frac{d}{dt}\left[\sum_{i=1}^{n} \frac{\partial L}{\partial \dot{q}_i}\dot{q}_i - L\right]$$

The term in brackets is the total energy of the system and is a constant of motion.

$$E = \frac{\partial L}{\partial \dot{\mathbf{q}}} \cdot \dot{\mathbf{q}} - L = \sum_{i=1}^{n} \frac{\partial L}{\partial \dot{q}_i}\dot{q}_i - L = \text{constant} \qquad (12.2.1)$$

We directly prove conservation of energy for Lagrangians of the form (12.1.3). Using formula (2.6.9), we first compute

$$\frac{\partial L}{\partial \dot{q}_i} = \sum_{j=1}^{n} A_{ij}(q_1, \ldots, q_n)\dot{q}_j$$

We can also write the vector form

$$\frac{\partial L}{\partial \dot{\mathbf{q}}} = \mathbf{A}(\mathbf{q})\dot{\mathbf{q}}$$

With this result, calculate the sum

$$\sum_{i=1}^{n} \frac{\partial L}{\partial \dot{q}_i}\dot{q}_i = \sum_{i=1}^{n}\sum_{j=1}^{n} A_{ij}(q_1, \ldots, q_n)\dot{q}_i\dot{q}_j = \dot{\mathbf{q}}^T \mathbf{A}(\mathbf{q})\dot{\mathbf{q}}$$

We now directly compute

$$\frac{d}{dt}\left[\sum_{i=1}^{n}\frac{\partial L}{\partial \dot{q}_i}\dot{q}_i - L\right] = 0$$

$$\frac{d}{dt}\left[\dot{\mathbf{q}}^T \mathbf{A}(\mathbf{q})\dot{\mathbf{q}} - \left[\frac{1}{2}\dot{\mathbf{q}}^T \mathbf{A}(\mathbf{q})\dot{\mathbf{q}} - U(\mathbf{q})\right]\right] = 0$$

$$\frac{d}{dt}\left[\frac{1}{2}\dot{\mathbf{q}}^T \mathbf{A}(\mathbf{q})\dot{\mathbf{q}} + U(\mathbf{q})\right] = 0$$

$$\frac{d}{dt}[T(\mathbf{q},\dot{\mathbf{q}}) + U(\mathbf{q})] = 0$$

This gives the conservation of energy.

$$E = \sum_{i=1}^{n}\frac{\partial L}{\partial \dot{q}_i}\dot{q}_i - L = T + U = \text{constant} \qquad (12.2.2)$$

12.3 Momentum

The generalized momenta p_i for a multiple DOF system is defined as

$$p_i = \frac{\partial L}{\partial \dot{q}_i}, \qquad i = 1,\ldots,n \qquad (12.3.1)$$

With Lagrange's equations (12.1.5), we have

$$\dot{p}_i = \frac{d}{dt}\frac{\partial L}{\partial \dot{q}_i} = \frac{\partial L}{\partial q_i}, \qquad i = 1,\ldots,n \qquad (12.3.2)$$

We can define $F_i = \dot{p}_i$ as a generalized force. If q_i does not appear in the Lagrangian, then its associated momentum p_i is conserved.

The vector form of momentum is written as

$$\mathbf{p} = \frac{\partial L}{\partial \dot{\mathbf{q}}}, \qquad \dot{\mathbf{p}} = \frac{\partial L}{\partial \mathbf{q}} \qquad (12.3.3)$$

12.4 Hamiltonian Dynamics

The derivation of Hamilton's equations of motion are provided in this section. This is accomplished using the *Legendre transformation*. We first compute the total differential of the most general form of the Lagrangian $L(\mathbf{q}, \dot{\mathbf{q}}, t)$.

$$dL = \sum_{i=1}^{n} \frac{\partial L}{\partial q_i} dq_i + \sum_{i=1}^{n} \frac{\partial L}{\partial \dot{q}_i} d\dot{q} + \frac{\partial L}{\partial t} dt \qquad (12.4.1)$$

We use equations (12.3.1) and (12.3.2) to write

$$dL = \sum_{i=1}^{n} \dot{p}_i \, dq_i + \sum_{i=1}^{n} p_i \, d\dot{q} + \frac{\partial L}{\partial t} dt \qquad (12.4.2)$$

We also note the product rule for total differentials

$$d\left(\sum_{i=1}^{n} p_i \dot{q}_i\right) = \sum_{i=1}^{n} \dot{q}_i \, dp_i + \sum_{i=1}^{n} p_i \, d\dot{q}_i$$

$$d\left(\sum_{i=1}^{n} p_i \dot{q}_i\right) - \sum_{i=1}^{n} \dot{q}_i \, dp_i = \sum_{i=1}^{n} p_i \, d\dot{q}_i$$

This equation is substituted into equation (12.4.2).

$$dL = \sum_{i=1}^{n} \dot{p}_i \, dq_i + d\left(\sum_{i=1}^{n} p_i \dot{q}_i\right) - \sum_{i=1}^{n} \dot{q}_i \, dp_i + \frac{\partial L}{\partial t} dt$$

$$-dL = -\sum_{i=1}^{n} \dot{p}_i \, dq_i - d\left(\sum_{i=1}^{n} p_i \dot{q}_i\right) + \sum_{i=1}^{n} \dot{q}_i \, dp_i - \frac{\partial L}{\partial t} dt$$

$$d\left(\sum_{i=1}^{n} p_i \dot{q}_i\right) - dL = -\sum_{i=1}^{n} \dot{p}_i \, dq_i + \sum_{i=1}^{n} \dot{q}_i \, dp_i - \frac{\partial L}{\partial t} dt$$

$$d\left[\sum_{i=1}^{n} p_i \dot{q}_i - L\right] = -\sum_{i=1}^{n} \dot{p}_i \, dq_i + \sum_{i=1}^{n} \dot{q}_i \, dp_i - \frac{\partial L}{\partial t} dt$$

The term in brackets on the left hand side is the definition of the Hamiltonian H.

$$H = \sum_{i=1}^{n} p_i \dot{q}_i - L$$

12.4. HAMILTONIAN DYNAMICS

If the previous expressions define perfect differentials, then we observe

$$d\left[\sum_{i=1}^{n} p_i \dot{q}_i - L\right] = -\sum_{i=1}^{n} \dot{p}_i \, dq_i + \sum_{i=1}^{n} \dot{q}_i \, dp_i - \frac{\partial L}{\partial t} dt$$

$$dH = \sum_{i=1}^{n} \frac{\partial H}{\partial q_i} dq_i + \sum_{i=1}^{n} \frac{\partial H}{\partial p_i} dp_i + \frac{\partial H}{\partial t} dt$$

The differentials match if

$$\frac{\partial H}{\partial q_i} = -\dot{p}_i, \quad \frac{\partial H}{\partial p_i} = \dot{q}_i, \quad i = 1, \ldots, n \qquad \frac{\partial H}{\partial t} = -\frac{\partial L}{\partial t}$$

We have now obtained a new way of describing the dynamics of the system.

Definition 12.4.1 (Hamiltonian). The *Hamiltonian* $H(\mathbf{q}, \mathbf{p}, t)$ of a multiple DOF system is defined in terms of the Lagrangian as

$$H = \sum_{i=1}^{n} p_i \dot{q}_i - L = \mathbf{p} \cdot \dot{\mathbf{q}} - L \qquad (12.4.3)$$

The Hamiltonian uses the generalized coordinates $\mathbf{q}$ and momenta $\mathbf{p}$. This is in contrast to the Lagrangian which uses generalized coordinates $\mathbf{q}$ and velocities $\dot{\mathbf{q}}$.

Theorem 12.4.1 (Hamilton's Equations). *The dynamical path $\mathbf{q}(t)$ and $\mathbf{p}(t)$ of a classical multiple DOF system is a solution to the differential equations*

$$\dot{q}_i = \frac{\partial H}{\partial p_i} \qquad (12.4.4)$$

$$\dot{p}_i = -\frac{\partial H}{\partial q_i} \qquad (12.4.5)$$

for $i = 1, \ldots, n$.

Hamilton's equations are a system of $2n$ 1st order ordinary differential equations. The initial position $\mathbf{q}_0$ and initial momenta $\mathbf{p}_0$ determine the $2n$ constants of integration.

The vector form of Hamilton's equations is

$$\dot{\mathbf{q}} = \frac{\partial H}{\partial \mathbf{p}} \qquad (12.4.6)$$

$$\dot{\mathbf{p}} = -\frac{\partial H}{\partial \mathbf{q}} \qquad (12.4.7)$$

Equation (12.4.3) formally defines the Hamiltonian H. However, H must be a function of $\mathbf{p}$ instead of $\dot{\mathbf{q}}$. The transformation begins by solving the following equations for $\dot{q}_i$.

$$p_i = \frac{\partial L}{\partial \dot{q}_i}, \quad i = 1, \ldots, n$$

We then obtain the functions: $\dot{q}_i = f_i(q_1, \ldots, q_n, p_1, \ldots p_n)$ for $i = 1, \ldots, n$. These equations are substituted into equation (12.4.3) to get the desired form of H.

In order to better understand the process of converting from the Lagrangian to the Hamiltonian, we perform the Legendre transformation on a specific example. Consider the Lagrangian of the form (12.1.3).

$$L(\mathbf{q}, \dot{\mathbf{q}}) = \frac{1}{2}\dot{\mathbf{q}}^T \mathbf{A}(\mathbf{q})\, \dot{\mathbf{q}} - U(\mathbf{q}) = \frac{1}{2} \sum_{i=1}^{n} \sum_{j=1}^{n} A_{ij}(q_1, \ldots, q_n)\, \dot{q}_i \dot{q}_j - U(q_1, \ldots, q_n)$$

The momenta p_i are defined by

$$p_i = \frac{\partial L}{\partial \dot{q}_i} = \sum_{j=1}^{n} A_{ij}(q_1, \ldots, q_n)\, \dot{q}_j$$

In matrix-vector form:

$$\mathbf{p} = \mathbf{A}(\mathbf{q})\, \dot{\mathbf{q}}$$

We invert this equation in order to express the generalized velocities in terms of momenta.

$$\dot{\mathbf{q}} = \mathbf{A}^{-1}(\mathbf{q})\, \mathbf{p}$$

The matrix $\mathbf{A}(\mathbf{q})$ is invertible because it is symmetric positive definite. The Lagrangian can now be written as

$$\begin{aligned}
L(\mathbf{q}, \mathbf{p}) &= \frac{1}{2} \left[\mathbf{A}^{-1}(\mathbf{q})\, \mathbf{p} \right]^T \mathbf{A}(\mathbf{q}) \left[\mathbf{A}^{-1}(\mathbf{q})\, \mathbf{p} \right] - U(\mathbf{q}) \\
&= \frac{1}{2} \mathbf{p}^T \left[\mathbf{A}^{-1}(\mathbf{q}) \right]^T \mathbf{A}(\mathbf{q})\, \mathbf{A}^{-1}(\mathbf{q})\, \mathbf{p} - U(\mathbf{q}) \\
&= \frac{1}{2} \mathbf{p}^T \left[\mathbf{A}^{-1}(\mathbf{q}) \right]^T \mathbf{p} - U(\mathbf{q}) \\
&= \frac{1}{2} \mathbf{p}^T \mathbf{A}^{-1}(\mathbf{q})\, \mathbf{p} - U(\mathbf{q})
\end{aligned}$$

The last step is due to the fact that the inverse of a symmetric matrix is symmetric.

12.4. HAMILTONIAN DYNAMICS

We compute the Hamiltonian in terms of $\mathbf{q}$ and $\mathbf{p}$ in the following steps:

$$\begin{aligned} H(\mathbf{q},\mathbf{p}) &= \mathbf{p}^T \dot{\mathbf{q}} - L(\mathbf{q},\dot{\mathbf{q}}) \\ &= \mathbf{p}^T \left[\mathbf{A}^{-1}(\mathbf{q})\,\mathbf{p} \right] - L(\mathbf{q},\mathbf{p}) \\ &= \mathbf{p}^T \mathbf{A}^{-1}(\mathbf{q})\,\mathbf{p} - \frac{1}{2}\mathbf{p}^T \mathbf{A}^{-1}(\mathbf{q})\,\mathbf{p} + U(\mathbf{q}) \\ &= \frac{1}{2}\mathbf{p}^T \mathbf{A}^{-1}(\mathbf{q})\,\mathbf{p} + U(\mathbf{q}) \end{aligned}$$

Thus we have

$$H(\mathbf{q},\mathbf{p}) = T(\mathbf{q},\mathbf{p}) + U(\mathbf{q}), \qquad T(\mathbf{q},\mathbf{p}) = \frac{1}{2}\mathbf{p}^T \mathbf{A}^{-1}(\mathbf{q})\,\mathbf{p} \qquad (12.4.8)$$

The Hamiltonian is the total energy of the system. Furthermore, it is constant because the system was shown to be conservative in equation (12.2.2).

It is instructive to compute the total time derivative of the general Hamiltonian $H(\mathbf{p},\mathbf{q},t)$.

$$\frac{dH}{dt} = \sum_{i=1}^{n} \frac{\partial H}{\partial q_i} \dot{q}_i + \sum_{i=1}^{n} \frac{\partial H}{\partial p_i} \dot{p}_i + \frac{\partial H}{\partial t}$$

Using Hamilton's equations (12.4.4) and (12.4.5):

$$\frac{dH}{dt} = -\sum_{i=1}^{n} \dot{p}_i \dot{q}_i + \sum_{i=1}^{n} \dot{q}_i \dot{p}_i + \frac{\partial H}{\partial t} = \frac{\partial H}{\partial t}$$

If the Hamiltonian is not an explicit function of time t, then the remaining partial derivative vanishes. We also note

$$\frac{dH}{dt} = \frac{d}{dt}\left[\sum_{i=1}^{n} p_i \dot{q}_i - L \right] = \frac{d}{dt}\left[\sum_{i=1}^{n} \frac{\partial L}{\partial \dot{q}_i} \dot{q}_i - L \right] = \frac{dE}{dt} = 0 \qquad (12.4.9)$$

This is the conservation of energy and the Hamiltonian is total energy when it does not contain time explicitly.

12.5 Action Revisited

In this section, we consider the action S as a function of time. Let $T = [t_0, \infty)$ be the times of interest. The solution and variational spaces are defined as

$$\mathcal{U} = \{\mathbf{q} \,|\, q_i \in C^1(T), \quad \mathbf{q}(t_0) = \mathbf{q}_0\}$$
$$\mathcal{V} = \{\delta\mathbf{q} \,|\, \delta q_i \in C^1(T), \quad \delta\mathbf{q}(t_0) = \mathbf{0}\}$$

Action is the functional $S : \mathcal{U} \times T \to \mathbb{R}$ defined by

$$S[\mathbf{q}](t) = \int_{t_0}^{t} L(\mathbf{q}, \dot{\mathbf{q}}, \tau) \, d\tau \qquad (12.5.1)$$

For convenience, let $S \equiv S[\mathbf{q}](t)$ and $L \equiv L(\mathbf{q}, \dot{\mathbf{q}}, t)$ and note that

$$\frac{dS}{dt} = L \qquad (12.5.2)$$

We fix t and consider S as a functional of the *actual dynamic path* $\mathbf{q}(t)$ of the system. Computing the 1st variation of S yields

$$\delta S = \sum_{i=1}^{n} \left[\frac{\partial L}{\partial \dot{q}_i} \delta q_i\right]\bigg|_{t_0}^{t} + \sum_{i=1}^{n} \int_{t_0}^{t} \left[\frac{\partial L}{\partial q_i} - \frac{d}{d\tau}\frac{\partial L}{\partial \dot{q}_i}\right] \delta q_i \, d\tau$$

The integrals vanish because the term in brackets are Lagrange's equations. No variation is permitted at t_0 because the initial conditions are fixed. However, we may have arbitrary variations at t. Setting $p_i \equiv p_i(t)$, $\delta q_i \equiv \delta q_i(t)$, and using the definition of momentum, we have

$$\delta S = \sum_{i=1}^{n} \frac{\partial L}{\partial \dot{q}_i}\bigg|_t \delta q_i(t) = \sum_{i=1}^{n} p_i(t) \, \delta q_i(t) = \sum_{i=1}^{n} p_i \, \delta q_i$$

On the other hand, the variation of S as a function of $\mathbf{q}$ is given by

$$\delta S = \sum_{i=1}^{n} \frac{\partial S}{\partial q_i} \delta q_i$$

By comparing the last two expressions, we conclude that

$$p_i = \frac{\partial S}{\partial q_i} \qquad (12.5.3)$$

The total differential and total time derivative of S are

$$dS = \sum_{i=1}^{n} \frac{\partial S}{\partial q_i} dq_i + \frac{\partial S}{\partial t} dt \qquad (12.5.4)$$

$$\frac{dS}{dt} = \sum_{i=1}^{n} \frac{\partial S}{\partial q_i} \dot{q}_i + \frac{\partial S}{\partial t} \qquad (12.5.5)$$

12.5. ACTION REVISITED

Combining equations (12.5.2), (12.5.3), (12.5.5) and using the definition (12.4.3) of H gives

$$\frac{\partial S}{\partial t} = L - \sum_{i=1}^{n} \frac{\partial S}{\partial q_i} \dot{q}_i = L - \sum_{i=1}^{n} p_i \dot{q}_i = -H \qquad (12.5.6)$$

Equation (12.5.6) can also be written as

$$\frac{\partial S}{\partial t} + H(\mathbf{q}, \mathbf{p}, t) = 0$$

Let $S \equiv S(\mathbf{q}, t) \equiv S[\mathbf{q}](t)$ and transform the momenta $\mathbf{p}$ using equation (12.5.3).

$$\frac{\partial S}{\partial t} + H\left(\mathbf{q}, \frac{\partial S}{\partial \mathbf{q}}, t\right) = 0, \qquad \mathbf{p} = \frac{\partial S}{\partial \mathbf{q}}$$

This gives us another important equation of motion.

Theorem 12.5.1 (Hamilton-Jacobi Equation). *The dynamical path* $\mathbf{q}(t)$ *of a classical multiple DOF system is a solution to the differential equation*

$$\frac{\partial S}{\partial t} + H\left(q_1, \ldots, q_n, \frac{\partial S}{\partial q_1}, \ldots, \frac{\partial S}{\partial q_n}, t\right) = 0 \qquad (12.5.7)$$

The Hamilton-Jacobi equation is a 1st order partial differential equation in the n generalized coordinates and time.

Hamilton's principle can be formulated in terms of the Hamiltonian. First, substitute the result of equation (12.5.6) into equation (12.5.4) and use (12.5.3) to get the differential

$$dS = \sum_{i=1}^{n} p_i \, dq_i - H \, dt \qquad (12.5.8)$$

This expression can be integrated to compute the action S.

$$S[\mathbf{q}, \mathbf{p}](t) = \int_{\mathbf{q}_0}^{\mathbf{q}} \mathbf{p} \cdot d\mathbf{q} - \int_{t_0}^{t} H(\mathbf{q}, \mathbf{p}, \tau) \, d\tau$$

We change the variable of integration for the first integral using the fact: $d\mathbf{q} = \dot{\mathbf{q}} \, dt$.

$$S[\mathbf{q}, \mathbf{p}](t) = \int_{t_0}^{t} \mathbf{p} \cdot \dot{\mathbf{q}} \, d\tau - \int_{t_0}^{t} H(\mathbf{q}, \mathbf{p}, \tau) \, d\tau$$

$$= \int_{t_0}^{t} [\mathbf{p} \cdot \dot{\mathbf{q}} - H(\mathbf{q}, \mathbf{p}, \tau)] \, d\tau$$

Now fix the values of $\mathbf{q}$ and $\mathbf{p}$ at the initial and final times

$$\mathbf{q}(t_0) = \mathbf{q}_0, \qquad \mathbf{p}(t_0) = \mathbf{p}_0, \qquad \mathbf{q}(t_f) = \mathbf{q}_f, \qquad \mathbf{p}(t_f) = \mathbf{p}_f$$

The value of S must be same regardless of formulation.

Theorem 12.5.2 (Hamilton's Principle). *The dynamical path $\mathbf{q}(t)$ and $\mathbf{p}(t)$ of a classical multiple DOF system from t_0 to t_f is a stationary point of the functional*

$$S[\mathbf{q},\mathbf{p}] = \int_{t_0}^{t_f} [\mathbf{p} \cdot \dot{\mathbf{q}} - H(\mathbf{q},\mathbf{p},t)]\, dt \tag{12.5.9}$$

In other words, $\delta S = 0$. The dynamical path $\mathbf{q}(t)$ and $\mathbf{p}(t)$ satisfies Hamilton's equations

$$\dot{\mathbf{q}} = \frac{\partial H}{\partial \mathbf{p}} \tag{12.5.10}$$

$$\dot{\mathbf{p}} = -\frac{\partial H}{\partial \mathbf{q}} \tag{12.5.11}$$

Proof. This is a restatement of Hamilton's principle with the Lagrangian replaced by the integrand in equation (12.5.9). We show that $\delta S = 0$ for this formulation produces Hamilton's equations. Computing δS and integrating by parts gives

$$\begin{aligned}
\delta S &= \int_{t_0}^{t_f} \left\{ \sum_{i=1}^{n} \left[\dot{q}_i\, \delta p_i + p_i\, \delta \dot{q}_i - \frac{\partial H}{\partial q_i} \delta q_i - \frac{\partial H}{\partial p_i} \delta p_i \right] \right\} dt \\
&= \sum_{i=1}^{n} \int_{t_0}^{t_f} \left[\dot{q}_i\, \delta p_i + p_i\, \delta \dot{q}_i - \frac{\partial H}{\partial q_i} \delta q_i - \frac{\partial H}{\partial p_i} \delta p_i \right] dt \\
&= \sum_{i=1}^{n} \int_{t_0}^{t_f} \left[\dot{q}_i\, \delta p_i - \dot{p}_i\, \delta q_i - \frac{\partial H}{\partial q_i} \delta q_i - \frac{\partial H}{\partial p_i} \delta p_i \right] dt + \sum_{i=1}^{n} p_i\, \delta q_i \big|_{t_0}^{t_f}
\end{aligned}$$

The boundary terms vanish because the variations are zero at the endpoints. We now set $\delta S = 0$.

$$\delta S = \sum_{i=1}^{n} \left\{ -\int_{t_0}^{t_f} \left[\dot{p}_i + \frac{\partial H}{\partial q_i} \right] \delta q_i\, dt + \int_{t_0}^{t_f} \left[\dot{q}_i - \frac{\partial H}{\partial p_i} \right] \delta p_i\, dt \right\} = 0$$

Each integral must vanish because the variations are independent. Applying the Fundamental Lemma 4.6.1 to each integral implies the terms in brackets are zero.

$$\dot{p}_i + \frac{\partial H}{\partial q_i} = 0, \qquad \dot{q}_i - \frac{\partial H}{\partial p_i} = 0, \qquad i = 1, \ldots, n$$

These are Hamilton's equations. □

12.6 Canonical Transformations

In some situations it may be necessary to switch variables. Suppose that we wish replace the old variables $\mathbf{q} = [q_1, \ldots, q_n]^T$ and $\mathbf{p} = [p_1, \ldots, p_n]^T$ with the new variables $\mathbf{Q} = [Q_1, \ldots, Q_n]^T$ and $\mathbf{P} = [P_1, \ldots, P_n]^T$. The change of variables is accomplished through

$$Q_i = f_i(\mathbf{q}, \mathbf{p}, t), \qquad P_i = g_i(\mathbf{q}, \mathbf{p}, t), \qquad i = 1, \ldots, n$$

Notice that the transformations may depend on time. Not all transformations of this type are useful or interesting. We require that the new variables obey Hamilton's equations of motion. Such transformations are called *canonical*.

Definition 12.6.1 (Canonical Transformations). Suppose that $\mathbf{q}(t)$ and $\mathbf{p}(t)$ are solutions to Hamilton's equations with Hamiltonian H. A transformation from these dynamic variables to a new set of dynamic variables $\mathbf{Q}(t)$ and $\mathbf{P}(t)$ is called a *canonical transformation* if the new dynamic variables obey Hamilton's equations for a new Hamiltonian $\tilde{H}$.

$$\dot{Q}_i = \frac{\partial \tilde{H}}{\partial P_i} \tag{12.6.1}$$

$$\dot{P}_i = -\frac{\partial \tilde{H}}{\partial Q_i} \tag{12.6.2}$$

for $i = 1, \ldots, n$. The Hamiltonians H and $\tilde{H}$ must describe the same physical system.

An important requirement concerning canonical transformations is that Hamilton's principle must be true in both sets of variables.

$$\delta \int_{t_0}^{t_f} [\mathbf{p} \cdot \dot{\mathbf{q}} - H(\mathbf{q}, \mathbf{p}, t)] \, dt = 0 \tag{12.6.3}$$

$$\delta \int_{t_0}^{t_f} \left[\mathbf{P} \cdot \dot{\mathbf{Q}} - \tilde{H}(\mathbf{Q}, \mathbf{P}, t) \right] dt = 0 \tag{12.6.4}$$

We need to show the equivalence of the integrands of both functionals. This is necessary in order to ensure that both sets of dynamic variables describe the same physical system. In doing so, we will establish a relationship between the old and new variables and the Hamiltonians H and $\tilde{H}$.

Lemma 12.6.1. *Suppose that $K \equiv \mathbf{P} \cdot \dot{\mathbf{Q}} - \tilde{H}(\mathbf{Q}, \mathbf{P}, t)$ is a function such that*

$$\delta \int_{t_0}^{t_f} K \, dt = 0$$

where no variation is allowed at the endpoints. Let G be the total time derivative of another function $F \equiv F(\mathbf{q}, \mathbf{p}, \mathbf{Q}, \mathbf{P}, t)$.

$$G = \frac{dF}{dt}$$

If no variation is allowed at the endpoints in any variable, then

$$\delta \int_{t_0}^{t_f} [K + G] \, dt = 0$$

Proof. The linearity properties of variation and integration are used to compute

$$\begin{aligned}
\delta \int_{t_0}^{t_f} [K + G] \, dt &= \delta \left[\int_{t_0}^{t_f} K \, dt + \int_{t_0}^{t_f} G \, dt \right] \\
&= \delta \int_{t_0}^{t_f} K \, dt + \delta \int_{t_0}^{t_f} G \, dt \\
&= 0 + \delta \int_{t_0}^{t_f} \frac{dF}{dt} \, dt \\
&= \delta \left[F \big|_{t_0}^{t_f} \right] \\
&= \sum_{i=1}^{n} \left[\frac{\partial F}{\partial q_i} \delta q_i + \frac{\partial F}{\partial p_i} \delta p_i + \frac{\partial F}{\partial Q_i} \delta Q_i + \frac{\partial F}{\partial P_i} \delta P_i \right] \bigg|_{t_0}^{t_f} \\
&= 0
\end{aligned}$$

The final result is 0 because the variations vanish at the endpoints. $\square$

This lemma provides a requirement for the equivalence of the integrands in equations (12.6.3) and (12.6.4). These integrands differ by a function G that is the total time derivative of a function F of old variables, new variables, and time.

$$\mathbf{p} \cdot \dot{\mathbf{q}} - H(\mathbf{q}, \mathbf{p}, t) = \mathbf{P} \cdot \dot{\mathbf{Q}} - \tilde{H}(\mathbf{Q}, \mathbf{P}, t) + \frac{d}{dt} F(\mathbf{q}, \mathbf{p}, \mathbf{Q}, \mathbf{P}, t), \quad \forall t \in [t_0, t_f]$$

We prefer to use the differential form (12.5.8) of this equation

$$\sum_{i=1}^{n} p_i \, dq_i - H \, dt = \sum_{i=1}^{n} P_i \, dQ_i - \tilde{H} \, dt + dF$$

12.6. CANONICAL TRANSFORMATIONS

We rearrange the previous equation.

$$dF = \sum_{i=1}^{n} p_i \, dq_i - \sum_{i=1}^{n} P_i \, dQ_i + (\tilde{H} - H) dt \qquad (12.6.5)$$

The function F is called the *generating function* of the canonical transformation. Equation (12.6.5) provides a way to transform the variables.

Suppose the generating function has the form

$$F \equiv F_1(\mathbf{q}, \mathbf{Q}, t) \qquad (12.6.6)$$

Then according to equation (12.6.5), we have

$$p_i = \frac{\partial F_1}{\partial q_i}, \quad P_i = -\frac{\partial F_1}{\partial Q_i}, \quad i = 1, \ldots, n, \qquad \tilde{H} = H + \frac{\partial F_1}{\partial t} \qquad (12.6.7)$$

Now consider the generating function

$$F \equiv F_2(\mathbf{q}, \mathbf{P}, t) - \mathbf{Q} \cdot \mathbf{P} \qquad (12.6.8)$$

We directly compare its total differential with the right hand side of equation (12.6.5).

$$dF = \sum_{i=1}^{n} \frac{\partial F_2}{\partial q_i} dq_i + \sum_{i=1}^{n} \frac{\partial F_2}{\partial P_i} dP_i + \frac{\partial F_2}{\partial t} dt - \sum_{i=1}^{n} P_i \, dQ_i - \sum_{i=1}^{n} Q_i \, dP_i$$

$$dF = \sum_{i=1}^{n} p_i \, dq_i - \sum_{i=1}^{n} P_i \, dQ_i + (\tilde{H} - H) dt$$

Setting the expressions equal and rearranging gives

$$\sum_{i=1}^{n} \frac{\partial F_2}{\partial q_i} dq_i + \sum_{i=1}^{n} \frac{\partial F_2}{\partial P_i} dP_i + \frac{\partial F_2}{\partial t} dt = \sum_{i=1}^{n} p_i \, dq_i + \sum_{i=1}^{n} Q_i \, dP_i + (\tilde{H} - H) dt$$

We then observe that

$$p_i = \frac{\partial F_2}{\partial q_i}, \quad Q_i = \frac{\partial F_2}{\partial P_i}, \quad i = 1, \ldots, n, \qquad \tilde{H} = H + \frac{\partial F_2}{\partial t} \qquad (12.6.9)$$

Consider the generating function

$$F \equiv F_3(\mathbf{p}, \mathbf{Q}, t) + \mathbf{q} \cdot \mathbf{p}$$

Compare its total differential with the right hand side of equation (12.6.5).

$$dF = \sum_{i=1}^{n} \frac{\partial F_3}{\partial p_i} dp_i + \sum_{i=1}^{n} \frac{\partial F_3}{\partial Q_i} dQ_i + \frac{\partial F_3}{\partial t} dt + \sum_{i=1}^{n} p_i \, dq_i + \sum_{i=1}^{n} q_i \, dp_i$$

$$dF = \sum_{i=1}^{n} p_i \, dq_i - \sum_{i=1}^{n} P_i \, dQ_i + (\tilde{H} - H) dt$$

Set the expressions equal and rearrange the terms.

$$\sum_{i=1}^{n} \frac{\partial F_3}{\partial p_i} dp_i + \sum_{i=1}^{n} \frac{\partial F_3}{\partial Q_i} dQ_i + \frac{\partial F_3}{\partial t} dt = -\sum_{i=1}^{n} q_i \, dp_i - \sum_{i=1}^{n} P_i \, dQ_i + (\tilde{H} - H) dt$$

We then observe that

$$q_i = -\frac{\partial F_3}{\partial p_i}, \quad P_i = -\frac{\partial F_3}{\partial Q_i}, \quad i = 1, \ldots, n, \qquad \tilde{H} = H + \frac{\partial F_3}{\partial t} \qquad (12.6.10)$$

Lastly, we consider

$$F \equiv F_4(\mathbf{p}, \mathbf{P}, t) + \mathbf{q} \cdot \mathbf{p} - \mathbf{Q} \cdot \mathbf{P} \qquad (12.6.11)$$

Again compare the total differential with the right hand side of equation (12.6.5).

$$dF = \sum_{i=1}^{n} \frac{\partial F_4}{\partial p_i} dp_i + \sum_{i=1}^{n} \frac{\partial F_4}{\partial P_i} dP_i + \frac{\partial F_4}{\partial t} dt$$
$$+ \sum_{i=1}^{n} p_i \, dq_i + \sum_{i=1}^{n} q_i \, dp_i - \sum_{i=1}^{n} P_i \, dQ_i - \sum_{i=1}^{n} Q_i \, dP_i$$

$$dF = \sum_{i=1}^{n} p_i \, dq_i - \sum_{i=1}^{n} P_i \, dQ_i + (\tilde{H} - H) dt$$

Setting the expressions equal and rearranging gives

$$\sum_{i=1}^{n} \frac{\partial F_4}{\partial p_i} dp_i + \sum_{i=1}^{n} \frac{\partial F_4}{\partial P_i} dP_i + \frac{\partial F_4}{\partial t} dt = -\sum_{i=1}^{n} q_i \, dp_i + \sum_{i=1}^{n} Q_i \, dP_i + (\tilde{H} - H) dt$$

We then observe that

$$q_i = -\frac{\partial F_4}{\partial p_i}, \quad Q_i = \frac{\partial F_4}{\partial P_i}, \quad i = 1, \ldots, n, \qquad \tilde{H} = H + \frac{\partial F_4}{\partial t} \qquad (12.6.12)$$

12.6. CANONICAL TRANSFORMATIONS

Type	F	1st var.	2nd var.	$\tilde{H}$
I	$F_1(\mathbf{q}, \mathbf{Q}, t)$	$\mathbf{p} = \frac{\partial F_1}{\partial \mathbf{q}}$	$\mathbf{P} = -\frac{\partial F_1}{\partial \mathbf{Q}}$	$H + \frac{\partial F_1}{\partial t}$
II	$F_2(\mathbf{q}, \mathbf{P}, t) - \mathbf{Q} \cdot \mathbf{P}$	$\mathbf{p} = \frac{\partial F_2}{\partial \mathbf{q}}$	$\mathbf{Q} = \frac{\partial F_2}{\partial \mathbf{P}}$	$H + \frac{\partial F_2}{\partial t}$
III	$F_3(\mathbf{p}, \mathbf{Q}, t) + \mathbf{q} \cdot \mathbf{p}$	$\mathbf{q} = -\frac{\partial F_3}{\partial \mathbf{p}}$	$\mathbf{P} = -\frac{\partial F_3}{\partial \mathbf{Q}}$	$H + \frac{\partial F_3}{\partial t}$
IV	$F_4(\mathbf{p}, \mathbf{P}, t) + \mathbf{q} \cdot \mathbf{p} - \mathbf{Q} \cdot \mathbf{P}$	$\mathbf{q} = -\frac{\partial F_4}{\partial \mathbf{p}}$	$\mathbf{Q} = \frac{\partial F_4}{\partial \mathbf{P}}$	$H + \frac{\partial F_4}{\partial t}$

Table 12.6.1: Some canonical transformations

The previously discussed canonical transformations are summarized in table 12.6.1. Other canonical transformations exist and are discussed in [10]. The canonical transformation from variables $\mathbf{q}$ and $\mathbf{p}$ to variables $\mathbf{Q}$ and $\mathbf{P}$ is denoted as

$$(\mathbf{q}, \mathbf{p}) \xrightarrow[F]{} (\mathbf{Q}, \mathbf{P}) \tag{12.6.13}$$

where F is the generating function for the transformation.

We provide some examples of canonical transformations. The *identity transformation* is the type II transformation defined as

$$\mathbf{Q} = \mathbf{q}, \qquad \mathbf{P} = \mathbf{p}, \qquad F_2 = \mathbf{q} \cdot \mathbf{P} \tag{12.6.14}$$

The *exchange transformation* is the type I canonical transformation defined by

$$\mathbf{Q} = \mathbf{p} \qquad \mathbf{P} = -\mathbf{q}, \qquad F_1 = \mathbf{q} \cdot \mathbf{Q} \tag{12.6.15}$$

This transformation exchanges the roles of the position and momentum variables.

The motion of the system is a canonical transformation.

$$\mathbf{Q} = \mathbf{q}(t + \tau), \qquad \mathbf{P} = \mathbf{p}(t + \tau), \qquad \tilde{H}(\mathbf{Q}, \mathbf{P}, t) - H(\mathbf{q}(t+\tau), \mathbf{p}(t+\tau), t+\tau) \tag{12.6.16}$$

where $\tau \geq 0$ is a parameter. The transformation can be proven to be canonical without a generating function. Instead, we directly compute

$$\delta \int_{t_0}^{t_f} \left[\mathbf{P} \cdot \dot{\mathbf{Q}} - \tilde{H}(\mathbf{Q}, \mathbf{P}, t) \right] dt = \delta \int_{t_0+\tau}^{t_f+\tau} \left[\mathbf{p} \cdot \dot{\mathbf{q}} - H(\mathbf{q}, \mathbf{p}, t) \right] dt = 0$$

The *point transformations* are type II canonical transformations.

$$p_i = \frac{\partial F_2}{\partial q_i}, \quad Q_i = f_i(\mathbf{q}, t), \quad i = 1, \ldots, n \qquad F_2 = \sum_{i=1}^{n} f_i(\mathbf{q}, t)\, P_i \qquad (12.6.17)$$

where $f_i(\mathbf{q}, t)$ are known functions. Solution for the new momenta P_i is obtained by using

$$p_i = \frac{\partial F_2}{\partial q_i} = \sum_{j=1}^{n} \left[\frac{\partial}{\partial q_i} f_j(\mathbf{q}, t) \right] P_j$$

This can be written in matrix form as

$$\mathbf{p} = \mathbf{F}(\mathbf{q}, t)\, \mathbf{P}, \qquad [\mathbf{F}(\mathbf{q}, t)]_{ij} = \frac{\partial}{\partial q_i} f_j(\mathbf{q}, t)$$

Assuming that the matrix $\mathbf{F}$ is invertible, we have

$$\mathbf{P} = \mathbf{F}^{-1}(\mathbf{q}, t)\, \mathbf{p}$$

The next theorem states that two successive canonical transformations form a canonical transformation.

Theorem 12.6.1. *Suppose we have the following set of canonical transformations*

$$(\mathbf{q}, \mathbf{p}) \xrightarrow[F]{} (\hat{\mathbf{Q}}, \hat{\mathbf{P}}), \qquad (\hat{\mathbf{Q}}, \hat{\mathbf{P}}) \xrightarrow[G]{} (\mathbf{Q}, \mathbf{P})$$

then the following transformation is canonical

$$(\mathbf{q}, \mathbf{p}) \xrightarrow[F+G]{} (\mathbf{Q}, \mathbf{P})$$

Proof. The set of canonical transformations obey the following equations:

$$dF = \sum_{i=1}^{n} p_i\, dq_i - \sum_{i=1}^{n} \hat{P}_i\, d\hat{Q}_i + (\hat{H} - H)\, dt$$

$$dG = \sum_{i=1}^{n} \hat{P}_i\, d\hat{Q}_i - \sum_{i=1}^{n} P_i\, dQ_i + (\tilde{H} - \hat{H})\, dt$$

Adding these equations gives

$$d(F + G) = dF + dG = \sum_{i=1}^{n} p_i\, dq_i - \sum_{i=1}^{n} P_i\, dQ_i + (\tilde{H} - H)\, dt$$

Thus $F + G$ is the generator for a canonical transformation from $(\mathbf{q}, \mathbf{p})$ to $(\mathbf{Q}, \mathbf{P})$. □

12.6. CANONICAL TRANSFORMATIONS

Theorem 12.6.2. *Canonical transformations form a group.*

$$(\mathbf{q},\mathbf{p}) \xrightarrow[F]{} (\hat{\mathbf{Q}},\hat{\mathbf{P}}), \qquad (\hat{\mathbf{Q}},\hat{\mathbf{P}}) \xrightarrow[G]{} (\tilde{\mathbf{Q}},\tilde{\mathbf{P}}), \qquad (\tilde{\mathbf{Q}},\tilde{\mathbf{P}}) \xrightarrow[K]{} (\mathbf{Q},\mathbf{P})$$

1. Two successive canonical transformations are canonical.

$$(\mathbf{q},\mathbf{p}) \xrightarrow[F+G]{} (\mathbf{Q},\mathbf{P})$$

2. The identity transformation is canonical.

$$(\mathbf{q},\mathbf{p}) \xrightarrow[I]{} (\mathbf{q},\mathbf{p})$$

3. The inverse transformation is canonical.

$$(\hat{\mathbf{Q}},\hat{\mathbf{P}}) \xrightarrow[-F]{} (\mathbf{q},\mathbf{p})$$

4. Three successive canonical transformations are associative

$$(\mathbf{q},\mathbf{p}) \xrightarrow[(F+G)+K]{} (\mathbf{Q},\mathbf{P})$$

$$(\mathbf{q},\mathbf{p}) \xrightarrow[F+(G+K)]{} (\mathbf{Q},\mathbf{P})$$

Proof. Property 1 was proven in the last theorem. Property 2 was demonstrated by the formulation of the identity transformation. Property 3 is easy to prove.

$$-dF = \sum_{i=1}^{n} \hat{P}_i\, d\hat{Q}_i - \sum_{i=1}^{n} p_i\, dq_i + (H - \hat{H})dt$$

This differential is associated with the generating function $-F$ and transforms $(\hat{\mathbf{Q}},\hat{\mathbf{P}})$ to $(\mathbf{q},\mathbf{p})$. Property 3 is obvious since differentials obey the associate law of addition.

$$(dF + dG) + dK = dF + (dG + dK)$$

Then we note the linear operator property of differentials.

$$d(F + G) + dK = dF + d(G + K)$$

This implies the generators obey: $(F+G)+K = F+(G+K)$. The proof is now complete. □

12.7 Poisson Brackets

Poisson brackets provide a useful way of presenting information in classical mechanics. They are extensively used in Hamiltonian dynamics. In the next section, a Taylor series method is presented in terms of Poisson brackets. We begin with their definition and essential properties.

Definition 12.7.1. Let $f \equiv f(\mathbf{q}, \mathbf{p}, t)$ and $g \equiv g(\mathbf{q}, \mathbf{p}, t)$. The *Poisson bracket* of f and g is defined as

$$[f, g] = \sum_{i=1}^n \left[\frac{\partial f}{\partial p_i} \frac{\partial g}{\partial q_i} - \frac{\partial f}{\partial q_i} \frac{\partial g}{\partial p_i} \right] = \frac{\partial f}{\partial \mathbf{p}} \cdot \frac{\partial g}{\partial \mathbf{q}} - \frac{\partial f}{\partial \mathbf{q}} \cdot \frac{\partial g}{\partial \mathbf{p}} \qquad (12.7.1)$$

There are several important properties of the Poisson bracket which we now prove.

Theorem 12.7.1. *The Poisson bracket obeys the properties listed below. Let* $f \equiv f(\mathbf{q}, \mathbf{p}, t)$, $g \equiv g(\mathbf{q}, \mathbf{p}, t)$, $h \equiv h(\mathbf{q}, \mathbf{p}, t)$, *and* c *be a constant.*

$$[f, g] = -[g, f] \qquad (12.7.2)$$
$$[f, c] = 0 \qquad (12.7.3)$$
$$[f + g, h] = [f, h] + [g, h] \qquad (12.7.4)$$
$$[fg, h] = f[g, h] + g[f, h] \qquad (12.7.5)$$
$$\frac{\partial}{\partial t}[f, g] = \left[\frac{\partial f}{\partial t}, g\right] + \left[f, \frac{\partial g}{\partial t}\right] \qquad (12.7.6)$$
$$[f, q_k] = \frac{\partial f}{\partial p_k} \qquad (12.7.7)$$
$$[f, p_k] = -\frac{\partial f}{\partial q_k} \qquad (12.7.8)$$

Proof. The proofs of most of these properties are easier to carry out when one uses the vector and dot product definition of Poisson bracket.

$$[f, g] = \frac{\partial f}{\partial \mathbf{p}} \cdot \frac{\partial g}{\partial \mathbf{q}} - \frac{\partial f}{\partial \mathbf{q}} \cdot \frac{\partial g}{\partial \mathbf{p}} = -\left[\frac{\partial g}{\partial \mathbf{p}} \cdot \frac{\partial f}{\partial \mathbf{q}} - \frac{\partial g}{\partial \mathbf{q}} \cdot \frac{\partial f}{\partial \mathbf{p}}\right] = -[g, f]$$

12.7. POISSON BRACKETS

$[f, c] = 0$ is obvious because the partial derivatives of c are all zero.

$$\begin{aligned}
[f + g, h] &= \frac{\partial(f+g)}{\partial \mathbf{p}} \cdot \frac{\partial h}{\partial \mathbf{q}} - \frac{\partial(f+g)}{\partial \mathbf{q}} \cdot \frac{\partial h}{\partial \mathbf{p}} \\
&= \left(\frac{\partial f}{\partial \mathbf{p}} + \frac{\partial g}{\partial \mathbf{p}}\right) \cdot \frac{\partial h}{\partial \mathbf{q}} - \left(\frac{\partial f}{\partial \mathbf{q}} + \frac{\partial g}{\partial \mathbf{q}}\right) \cdot \frac{\partial h}{\partial \mathbf{p}} \\
&= \frac{\partial f}{\partial \mathbf{p}} \cdot \frac{\partial h}{\partial \mathbf{q}} + \frac{\partial g}{\partial \mathbf{p}} \cdot \frac{\partial h}{\partial \mathbf{q}} - \frac{\partial f}{\partial \mathbf{q}} \cdot \frac{\partial h}{\partial \mathbf{p}} - \frac{\partial g}{\partial \mathbf{q}} \cdot \frac{\partial h}{\partial \mathbf{p}} \\
&= \left[\frac{\partial f}{\partial \mathbf{p}} \cdot \frac{\partial h}{\partial \mathbf{q}} - \frac{\partial f}{\partial \mathbf{q}} \cdot \frac{\partial h}{\partial \mathbf{p}}\right] + \left[\frac{\partial g}{\partial \mathbf{p}} \cdot \frac{\partial h}{\partial \mathbf{q}} - \frac{\partial g}{\partial \mathbf{q}} \cdot \frac{\partial h}{\partial \mathbf{p}}\right] \\
&= [f, h] + [g, h]
\end{aligned}$$

$$\begin{aligned}
[fg, h] &= \frac{\partial(fg)}{\partial \mathbf{p}} \cdot \frac{\partial h}{\partial \mathbf{q}} - \frac{\partial(fg)}{\partial \mathbf{q}} \cdot \frac{\partial h}{\partial \mathbf{p}} \\
&= \left(\frac{\partial f}{\partial \mathbf{p}} g + \frac{\partial g}{\partial \mathbf{p}} f\right) \cdot \frac{\partial h}{\partial \mathbf{q}} - \left(\frac{\partial f}{\partial \mathbf{q}} g + \frac{\partial g}{\partial \mathbf{q}} f\right) \cdot \frac{\partial h}{\partial \mathbf{p}} \\
&= g \frac{\partial f}{\partial \mathbf{p}} \cdot \frac{\partial h}{\partial \mathbf{q}} + f \frac{\partial g}{\partial \mathbf{p}} \cdot \frac{\partial h}{\partial \mathbf{q}} - g \frac{\partial f}{\partial \mathbf{q}} \cdot \frac{\partial h}{\partial \mathbf{p}} - f \frac{\partial g}{\partial \mathbf{q}} \cdot \frac{\partial h}{\partial \mathbf{p}} \\
&= f \left[\frac{\partial g}{\partial \mathbf{p}} \cdot \frac{\partial h}{\partial \mathbf{q}} - \frac{\partial g}{\partial \mathbf{q}} \cdot \frac{\partial h}{\partial \mathbf{p}}\right] + g \left[\frac{\partial f}{\partial \mathbf{p}} \cdot \frac{\partial h}{\partial \mathbf{q}} - \frac{\partial f}{\partial \mathbf{q}} \cdot \frac{\partial h}{\partial \mathbf{p}}\right] \\
&= f[g, h] + g[f, h]
\end{aligned}$$

$$\begin{aligned}
\frac{\partial}{\partial t}[f, g] &= \frac{\partial}{\partial t}\left[\frac{\partial f}{\partial \mathbf{p}} \cdot \frac{\partial g}{\partial \mathbf{q}} - \frac{\partial f}{\partial \mathbf{q}} \cdot \frac{\partial g}{\partial \mathbf{p}}\right] \\
&= \left(\frac{\partial}{\partial t}\frac{\partial f}{\partial \mathbf{p}}\right) \cdot \frac{\partial g}{\partial \mathbf{q}} + \frac{\partial f}{\partial \mathbf{p}} \cdot \left(\frac{\partial}{\partial t}\frac{\partial g}{\partial \mathbf{q}}\right) - \left(\frac{\partial}{\partial t}\frac{\partial f}{\partial \mathbf{q}}\right) \cdot \frac{\partial g}{\partial \mathbf{p}} - \frac{\partial f}{\partial \mathbf{q}} \cdot \left(\frac{\partial}{\partial t}\frac{\partial g}{\partial \mathbf{p}}\right) \\
&= \left(\frac{\partial}{\partial t}\frac{\partial f}{\partial \mathbf{p}}\right) \cdot \frac{\partial g}{\partial \mathbf{q}} - \left(\frac{\partial}{\partial t}\frac{\partial f}{\partial \mathbf{q}}\right) \cdot \frac{\partial g}{\partial \mathbf{p}} + \frac{\partial f}{\partial \mathbf{p}} \cdot \left(\frac{\partial}{\partial t}\frac{\partial g}{\partial \mathbf{q}}\right) - \frac{\partial f}{\partial \mathbf{q}} \cdot \left(\frac{\partial}{\partial t}\frac{\partial g}{\partial \mathbf{p}}\right) \\
&= \left[\left(\frac{\partial}{\partial \mathbf{p}}\frac{\partial f}{\partial t}\right) \cdot \frac{\partial g}{\partial \mathbf{q}} - \left(\frac{\partial}{\partial \mathbf{q}}\frac{\partial f}{\partial t}\right) \cdot \frac{\partial g}{\partial \mathbf{p}}\right] \\
&\quad + \left[\frac{\partial f}{\partial \mathbf{p}} \cdot \left(\frac{\partial}{\partial \mathbf{q}}\frac{\partial g}{\partial t}\right) - \frac{\partial f}{\partial \mathbf{q}} \cdot \left(\frac{\partial}{\partial \mathbf{p}}\frac{\partial g}{\partial t}\right)\right] \\
&= \left[\frac{\partial f}{\partial t}, g\right] + \left[f, \frac{\partial g}{\partial t}\right]
\end{aligned}$$

The last two properties are proved using the summation notation of the Poisson bracket.

$$[f, q_k] = \sum_{i=1}^{n} \left[\frac{\partial f}{\partial p_i} \frac{\partial q_k}{\partial q_i} - \frac{\partial f}{\partial q_i} \frac{\partial q_k}{\partial p_i} \right] = \frac{\partial f}{\partial p_k}$$

$$[f, p_k] = \sum_{i=1}^{n} \left[\frac{\partial f}{\partial p_i} \frac{\partial p_k}{\partial q_i} - \frac{\partial f}{\partial q_i} \frac{\partial p_k}{\partial p_i} \right] = -\frac{\partial f}{\partial q_k}$$

□

Equations (12.7.7) and (12.7.8) (proved above) give the obvious relations

$$[q_i, q_j] = 0, \qquad [p_i, p_j] = 0, \qquad [p_i, q_j] = \delta_{ij}, \qquad i, j \in \{1, \ldots, n\} \qquad (12.7.9)$$

where δ_{ij} is the Kronecker delta function.

Using equations (12.7.7) and (12.7.8) with $f = H$, we recover Hamilton's equations

$$\dot{q}_i = \frac{\partial H}{\partial p_i} = [H, q_i], \qquad \dot{p}_i = -\frac{\partial H}{\partial q_i} = [H, p_i], \qquad i = 1, \ldots, n \qquad (12.7.10)$$

Theorem 12.7.2 (Jacobi Identity). *Let $f \equiv f(\mathbf{q}, \mathbf{p}, t)$, $g \equiv g(\mathbf{q}, \mathbf{p}, t)$, and $h \equiv h(\mathbf{q}, \mathbf{p}, t)$. The Poisson bracket obeys the property*

$$[f, [g, h]] + [g, [h, f]] + [h, [f, g]] = 0 \qquad (12.7.11)$$

The proof of the Jacobi identity is complicated and will be omitted. However, the interested reader will find the proof in [10, 22].

Computing the total time derivative of $f \equiv f(\mathbf{q}, \mathbf{p}, t)$, we have

$$\frac{df}{dt} = \frac{\partial f}{\partial t} + \sum_{i=1}^{n} \left[\frac{\partial f}{\partial q_i} \dot{q}_i + \frac{\partial f}{\partial p_i} \dot{p}_i \right]$$

Let H be the Hamiltonian of a system. Using Hamilton's equations, we can rewrite the above equation as

$$\frac{df}{dt} = \frac{\partial f}{\partial t} + \sum_{i=1}^{n} \left[\frac{\partial H}{\partial p_i} \frac{\partial f}{\partial q_i} - \frac{\partial H}{\partial q_i} \frac{\partial f}{\partial p_i} \right]$$

Using Poisson brackets, we now write

$$\frac{df}{dt} = \frac{\partial f}{\partial t} + [H, f] \qquad (12.7.12)$$

12.7. POISSON BRACKETS

If f is a constant of motion of the system, then we have

$$\frac{df}{dt} = \frac{\partial f}{\partial t} + [H, f] = 0 \qquad (12.7.13)$$

If the constant of motion is not an explicit function of time, then

$$[H, f] = 0 \qquad (12.7.14)$$

The next theorem allows us to search for constants of motion.

Theorem 12.7.3 (Poisson's Theorem). *If $f \equiv f(\mathbf{q}, \mathbf{p}, t)$ and $g \equiv g(\mathbf{q}, \mathbf{p}, t)$ are constants of motion of a system, then $[f, g]$ is also a constant of motion.*

$$[f, g] = constant \qquad (12.7.15)$$

Proof. Compute the total time derivative of equation (12.7.15) using equation (12.7.12).

$$\frac{d}{dt}[f, g] = \frac{\partial}{\partial t}[f, g] + [H, [f, g]]$$

Using equations (12.7.6) and (12.7.11) gives

$$\frac{d}{dt}[f, g] = \left[\frac{\partial f}{\partial t}, g\right] + \left[f, \frac{\partial g}{\partial t}\right] - [f, [g, H]] - [g, [H, f]]$$

Equation (12.7.2) is applied to the last term.

$$\frac{d}{dt}[f, g] = \left[\frac{\partial f}{\partial t}, g\right] + \left[f, \frac{\partial g}{\partial t}\right] - [f, [g, H]] + [[H, f], g]$$

Equation (12.7.4) allows us to combine brackets.

$$\frac{d}{dt}[f, g] = \left[\frac{\partial f}{\partial t} + [H, f], g\right] + \left[f, \frac{\partial g}{\partial t} - [g, H]\right]$$

Using (12.7.2) again inside the last bracket gives

$$\frac{d}{dt}[f, g] = \left[\frac{\partial f}{\partial t} + [H, f], g\right] + \left[f, \frac{\partial g}{\partial t} + [H, g]\right]$$

Finally, we have total derivatives inside each bracket according to equation (12.7.12).

$$\frac{d}{dt}[f, g] = \left[\frac{df}{dt}, g\right] + \left[f, \frac{dg}{dt}\right] = [0, g] + [f, 0] = 0$$

Thus $[f, g]$ is constant. $\square$

Unfortunately, Poisson's theorem does not always furnish a new independent constant of motion. Often the result is simply a constant or a function of f or g. If neither of these situations occur, then the resulting expression is a new constant of motion.

12.8 Infinitesimal Canonical Transformations

There are occasions when we wish to transform the dynamic variables by a small amount. This is expressed by the equations

$$\mathbf{Q} = \mathbf{q} + \Delta \mathbf{q}, \qquad \mathbf{P} = \mathbf{p} + \Delta \mathbf{p} \qquad (12.8.1)$$

This is a type II canonical transformation with the generating function

$$F = F_2 + \mathbf{Q} \cdot \mathbf{P}, \qquad F_2 = \mathbf{q} \cdot \mathbf{P} + \epsilon G(\mathbf{q}, \mathbf{P}, t)$$

where ϵ is a small parameter and G is any function whose 1st partial derivatives are continuous. From table 12.6.1, we have the relations

$$\mathbf{p} = \frac{\partial F_2}{\partial \mathbf{q}} = \mathbf{P} + \epsilon \frac{\partial G}{\partial \mathbf{q}}$$

$$\mathbf{Q} = \frac{\partial F_2}{\partial \mathbf{P}} = \mathbf{q} + \epsilon \frac{\partial G}{\partial \mathbf{P}}$$

These equations give

$$\Delta \mathbf{q} = \mathbf{Q} - \mathbf{q} = \epsilon \frac{\partial G}{\partial \mathbf{P}}, \qquad \Delta \mathbf{p} = \mathbf{P} - \mathbf{p} = -\epsilon \frac{\partial G}{\partial \mathbf{q}} \qquad (12.8.2)$$

We wish to evaluate the partial derivative of G with respect to $\mathbf{p}$ instead of $\mathbf{P}$. Fix the values of $\mathbf{q}$ and t. Then compute the i-th partial derivative using chain rule.

$$\left.\frac{\partial G}{\partial P_i}\right|_{\mathbf{P}} = \left[\left.\frac{\partial G}{\partial p_i}\right|_{\mathbf{P}}\right]\left[\left.\frac{\partial p_i}{\partial P_i}\right|_{\mathbf{P}}\right] = \left.\frac{\partial G}{\partial p_i}\right|_{\mathbf{P}}$$

where each derivative is evaluated at $\mathbf{P}$. We used the fact that $\partial p_i / \partial P_i = 1$. We write the 1st order Taylor series expansion

$$\left.\frac{\partial G}{\partial p_i}\right|_{\mathbf{P}} = \left.\frac{\partial G}{\partial p_i}\right|_{\mathbf{p}} + \sum_{j=1}^{n} \left.\frac{\partial^2 G}{\partial p_j \partial p_i}\right|_{\mathbf{p}} \Delta p_j + r_i(\Delta \mathbf{p})$$

where $r_i(\Delta \mathbf{p})$ is $o(\|\Delta \mathbf{p}\|_2)$ as $\Delta \mathbf{p} \to \mathbf{0}$. We argue that since $\Delta \mathbf{p}$ is small, the following approximation is valid:

$$\left.\frac{\partial G}{\partial P_i}\right|_{\mathbf{P}} = \left.\frac{\partial G}{\partial p_i}\right|_{\mathbf{P}} \approx \left.\frac{\partial G}{\partial p_i}\right|_{\mathbf{p}}$$

Henceforth, we shall use $G \equiv G(\mathbf{q}, \mathbf{p}, t)$ with

$$\Delta \mathbf{q} = \mathbf{Q} - \mathbf{q} = \epsilon \frac{\partial G}{\partial \mathbf{p}}, \qquad \Delta \mathbf{p} = \mathbf{P} - \mathbf{p} = -\epsilon \frac{\partial G}{\partial \mathbf{q}} \qquad (12.8.3)$$

12.8. INFINITESIMAL CANONICAL TRANSFORMATIONS

Let $G = H$, $\epsilon = \Delta t$, and divide both sides of equation (12.8.3) by Δt.

$$\frac{\Delta \mathbf{q}}{\Delta t} = \frac{\partial H}{\partial \mathbf{p}}, \qquad \frac{\Delta \mathbf{p}}{\Delta t} = -\frac{\partial H}{\partial \mathbf{q}}$$

Letting $\Delta t \to 0$ reproduces Hamilton's equations of motion.

$$\dot{\mathbf{q}} = \frac{\partial H}{\partial \mathbf{p}}, \qquad \dot{\mathbf{p}} = -\frac{\partial H}{\partial \mathbf{q}} \qquad (12.8.4)$$

Consider a function $u(\mathbf{q}, \mathbf{p})$ which depends on the dynamic variables $\mathbf{q}$ and $\mathbf{p}$. The 1st order Taylor series expansion of u is

$$u(\mathbf{q} + \Delta \mathbf{q}, \mathbf{p} + \Delta \mathbf{p}) = u(\mathbf{q}, \mathbf{p}) + \frac{\partial u}{\partial \mathbf{q}} \cdot \Delta \mathbf{q} + \frac{\partial u}{\partial \mathbf{p}} \cdot \Delta \mathbf{p} + r(\Delta \mathbf{q}, \Delta \mathbf{p}) \qquad (12.8.5)$$

Let $\Delta \mathbf{z} = [(\Delta \mathbf{q})^T, (\Delta \mathbf{p})^T]^T$. Then the error term $r(\Delta \mathbf{q}, \Delta \mathbf{p})$ is $o(\|\Delta \mathbf{z}\|_2)$ as $\Delta \mathbf{z} \to \mathbf{0}$. When $\Delta \mathbf{q}$ and $\Delta \mathbf{p}$ are small, we note the 1st order approximation

$$\Delta u = u(\mathbf{q} + \Delta \mathbf{q}, \mathbf{p} + \Delta \mathbf{p}) - u(\mathbf{q}, \mathbf{p}) \approx \frac{\partial u}{\partial \mathbf{q}} \cdot \Delta \mathbf{q} + \frac{\partial u}{\partial \mathbf{p}} \cdot \Delta \mathbf{p}$$

We accept this approximation as sufficiently accurate and simply write

$$\Delta u = \frac{\partial u}{\partial \mathbf{q}} \cdot \Delta \mathbf{q} + \frac{\partial u}{\partial \mathbf{p}} \cdot \Delta \mathbf{p} \qquad (12.8.6)$$

Now suppose that the increments $\Delta \mathbf{q}$ and $\Delta \mathbf{p}$ are given by an infinitesimal canonical transformation. Using equation (12.8.3), we write

$$\Delta \mathbf{q} = \alpha \frac{\partial G}{\partial \mathbf{p}}, \qquad \Delta \mathbf{p} = -\alpha \frac{\partial G}{\partial \mathbf{q}} \qquad (12.8.7)$$

where α is a small parameter. Substitution of equation (12.8.7) into equation (12.8.6) gives the following result:

$$\begin{aligned}
\Delta u &= \alpha \frac{\partial u}{\partial \mathbf{q}} \frac{\partial G}{\partial \mathbf{p}} - \alpha \frac{\partial u}{\partial \mathbf{p}} \frac{\partial G}{\partial \mathbf{q}} \\
&= \alpha \left[\frac{\partial u}{\partial \mathbf{q}} \frac{\partial G}{\partial \mathbf{p}} - \frac{\partial u}{\partial \mathbf{p}} \frac{\partial G}{\partial \mathbf{q}} \right] \\
&= \alpha \left[\frac{\partial G}{\partial \mathbf{p}} \frac{\partial u}{\partial \mathbf{q}} - \frac{\partial G}{\partial \mathbf{q}} \frac{\partial u}{\partial \mathbf{p}} \right]
\end{aligned}$$

The last line on the previous page can be written as a Poisson bracket.

$$\Delta u = \alpha [G, u] \tag{12.8.8}$$

Dividing by α and letting $\alpha \to 0$, we have the initial value problem

$$\frac{du}{d\alpha} = [G, u], \quad u(0) = u_0 \tag{12.8.9}$$

A Taylor series expansion of u in terms of α is

$$u(\alpha) = u_0 + \alpha \left.\frac{du}{d\alpha}\right|_{\alpha=0} + \frac{\alpha^2}{2} \left.\frac{d^2 u}{d\alpha^2}\right|_{\alpha=0} + \frac{\alpha^3}{3!} \left.\frac{d^3 u}{d\alpha^3}\right|_{\alpha=0} + \cdots$$

This expansion assumes that u is analytic at $\alpha = 0$. Replacing u in equation (12.8.9) with $du/d\alpha$ gives the expression

$$\frac{d^2 u}{d\alpha^2} = \left[G, \frac{du}{d\alpha}\right] = [G, [G, u]]$$

Higher order derivatives are computed recursively using the following formula:

$$\frac{d^n u}{d\alpha^n} = \left[G, \frac{d^{n-1} u}{d\alpha^{n-1}}\right] = [G, [\cdots [G, u]]]$$

The Taylor series can be written in terms of repeated Poisson brackets.

$$u(\alpha) = u_0 + \alpha \,[G, u]|_{\alpha=0} + \frac{\alpha^2}{2}\,[G, [G, u]]|_{\alpha=0} + \frac{\alpha^3}{3!}\,[G, [G, [G, u]]]|_{\alpha=0} + \cdots \tag{12.8.10}$$

Suppose that $G = H$ where H is the Hamiltonian of a physical system. Also, let $\alpha = t$ where t is time. Then we have the formal solution

$$u(t) = u_0 + t\,[H, u]|_{t=0} + \frac{t^2}{2}\,[H, [H, u]]|_{t=0} + \frac{t^3}{3!}\,[H, [H, [H, u]]]|_{t=0} + \cdots \tag{12.8.11}$$

Equation (12.8.11) gives the value of $u(t)$ during the actual motion of the system.

12.8. INFINITESIMAL CANONICAL TRANSFORMATIONS

This page intentionally left blank.

We demonstrate the Taylor series method for the single DOF mass-spring system with the following initial conditions:

$$H = \frac{p^2}{2m} + \frac{kx^2}{2}, \qquad x(0) = x_0, \quad p(0) = 0 \tag{12.8.12}$$

The solution for $x(t)$ is computed using equation (12.8.11).

$$x(t) = x_0 + t\,[H,x]|_{t=0} + \frac{t^2}{2}\,[H,[H,x]]|_{t=0} + \frac{t^3}{3!}\,[H,[H,[H,x]]]|_{t=0} + \cdots \tag{12.8.13}$$

Calculating the necessary partial derivatives of H gives

$$\frac{\partial H}{\partial p} = \frac{p}{m}, \qquad \frac{\partial H}{\partial x} = kx$$

The successive Poisson brackets are calculated as follows:

$$[H,x] = \frac{\partial H}{\partial p}\frac{\partial x}{\partial x} - \frac{\partial H}{\partial x}\frac{\partial x}{\partial p}$$
$$= \frac{p}{m} \cdot 1 - kx \cdot 0$$
$$= \frac{p}{m}$$

$$[H,x]|_{t=0} = \left.\frac{p}{m}\right|_{t=0} = 0$$

$$[H,[H,x]] = \frac{\partial H}{\partial p}\frac{\partial}{\partial x}\left(\frac{p}{m}\right) - \frac{\partial H}{\partial x}\frac{\partial}{\partial p}\left(\frac{p}{m}\right)$$
$$= \frac{p}{m} \cdot 0 - kx \cdot \frac{1}{m}$$
$$= -\frac{kx}{m}$$

$$[H,[H,x]]|_{t=0} = \left.-\frac{kx}{m}\right|_{t=0} = -\frac{kx_0}{m}$$

$$[H,[H,[H,x]]] = \frac{\partial H}{\partial p}\frac{\partial}{\partial x}\left(-\frac{kx}{m}\right) - \frac{\partial H}{\partial x}\frac{\partial}{\partial p}\left(-\frac{kx}{m}\right)$$
$$= \frac{p}{m} \cdot \left(-\frac{k}{m}\right) - kx \cdot 0$$
$$= -\frac{pk}{m^2}$$

$$[H,[H,[H,x]]]|_{t=0} = \left.-\frac{pk}{m^2}\right|_{t=0} = 0$$

12.8. INFINITESIMAL CANONICAL TRANSFORMATIONS

$$[H,[H,[H,[H,x]]]] = \frac{\partial H}{\partial p}\frac{\partial}{\partial x}\left(-\frac{pk}{m^2}\right) - \frac{\partial H}{\partial x}\frac{\partial}{\partial p}\left(-\frac{pk}{m^2}\right)$$

$$= \frac{p}{m}\cdot 0 - kx\cdot\left(-\frac{k}{m^2}\right)$$

$$= \frac{k^2 x}{m^2}$$

$$[H,[H,[H,[H,x]]]]|_{t=0} = \left.\frac{k^2 x}{m^2}\right|_{t=0} = \frac{k^2 x_0}{m^2}$$

Odd numbers of Poisson brackets are always 0. Inserting the above expressions into equation (12.8.13) produces

$$x(t) = x_0\left[1 - \frac{t^2}{2}\left(\frac{k}{m}\right) + \frac{t^4}{4!}\left(\frac{k}{m}\right)^2 - \cdots\right]$$

Letting $\omega^2 = k/m$, we have

$$x(t) = x_0\left[1 - \frac{1}{2}(\omega t)^2 + \frac{1}{4!}(\omega t)^4 - \cdots\right]$$

The term in brackets is recognized as the Taylor series for $\cos(\omega t)$.

$$x(t) = x_0 \cos(\omega t) \qquad (12.8.14)$$

This is indeed the known solution for the given initial conditions.

12.9 Central Force Problem

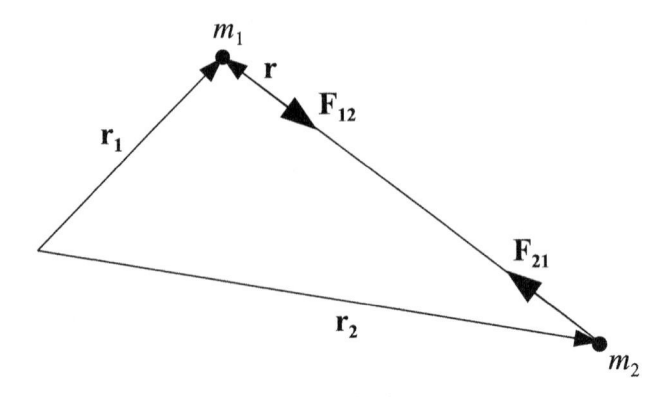

Figure 12.9.1: Central force problem

An important problem in classical mechanics is that of two bodies with a mutual force acting along the direction of their separation. The bodies are considered small compared to their separation distance. Thus, they are considered as point particles.

Consider two bodies with masses m_1 and m_2 shown in figure 12.9.1. Their position vectors in cartesian coordinates are $\mathbf{r}_1$ and $\mathbf{r}_2$ respectively. Let $\mathbf{r} = \mathbf{r}_1 - \mathbf{r}_2$ be the vector which connects the two masses. A attractive conservative force acts on each particle and obeys Newton's 3rd law. This force is derived from a potential energy function $U(r)$ which only depends on the separation distance $r = \|\mathbf{r}\|_2$.

$$\mathbf{F}_{12} = -\frac{\mathbf{r}}{r}U'(r), \qquad \mathbf{F}_{21} = \frac{\mathbf{r}}{r}U'(r)$$

In order to simplify matters, choose the center of mass to be the origin of a new cartesian coordinate system. The center of mass moves with constant velocity $\mathbf{V}$ because there are no external forces acting on the system. Hence, the center of mass system is an inertial frame in which we may formulate our problem. The location of the center of mass $\mathbf{R}$ is given by

$$\mathbf{R} = \frac{m_1 \mathbf{r}_1 + m_2 \mathbf{r}_2}{m_1 + m_2}$$

12.9. CENTRAL FORCE PROBLEM

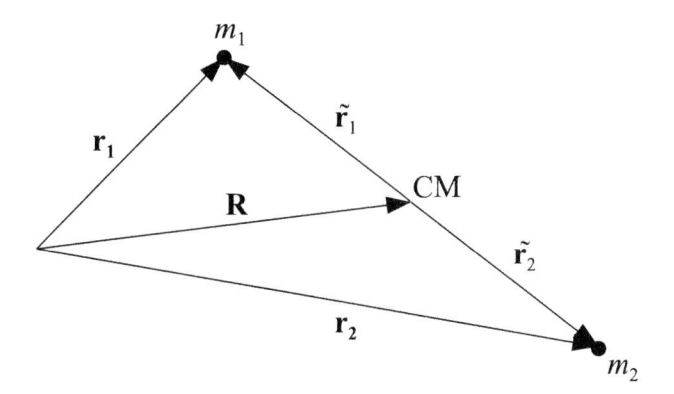

Figure 12.9.2: Center of mass coordinates

The transformation of coordinates is carried out through

$$\tilde{\mathbf{r}}_1 = \mathbf{r}_1 - \mathbf{R} = \frac{m_2}{m_1 + m_2} \mathbf{r}$$

$$\tilde{\mathbf{r}}_2 = \mathbf{r}_2 - \mathbf{R} = -\frac{m_1}{m_1 + m_2} \mathbf{r}$$

Figure 12.9.2 shows both coordinate systems.

Let $M = m_1 + m_2$ and notice that $\mathbf{r} = \tilde{\mathbf{r}}_1 - \tilde{\mathbf{r}}_2$. The Lagrangian, written in the center of mass coordinate system, is formulated as

$$L = \frac{1}{2} m_1 \left\| \dot{\tilde{\mathbf{r}}}_1 \right\|_2^2 + \frac{1}{2} m_2 \left\| \dot{\tilde{\mathbf{r}}}_2 \right\|_2^2 - U(r)$$

We simplify the Lagrangian in the following steps:

$$\begin{aligned} L &= \frac{1}{2} m_1 \left\| \frac{m_2}{M} \dot{\mathbf{r}} \right\|_2^2 + \frac{1}{2} m_2 \left\| -\frac{m_1}{M} \dot{\mathbf{r}} \right\|_2^2 - U(r) \\ &= \frac{m_1 m_2^2}{2M^2} \left\| \dot{\mathbf{r}} \right\|_2^2 + \frac{m_2 m_1^2}{2M^2} \left\| \dot{\mathbf{r}} \right\|_2^2 - U(r) \\ &= \frac{1}{2} \left[\frac{m_1 m_2 (m_2 + m_1)}{M^2} \right] \left\| \dot{\mathbf{r}} \right\|_2^2 - U(r) \\ &= \frac{1}{2} \left[\frac{m_1 m_2}{M} \right] \left\| \dot{\mathbf{r}} \right\|_2^2 - U(r) \end{aligned}$$

The Lagrangian can now be written in terms of the reduced mass μ.

$$L = \frac{1}{2} \mu \left\| \dot{\mathbf{r}} \right\|_2^2 - U(r), \qquad \mu = \frac{m_1 m_2}{m_1 + m_2} \qquad (12.9.1)$$

There are no external torques acting on the system and thus angular momentum **L** is conserved. This implies that the motion takes place in a plane normal to the angular momentum vector **L**. The problem is now 2-dimensional and we write the Lagrangian in polar coordinates.

$$L = \frac{1}{2}\mu\left[\dot{r}^2 + r^2\dot{\theta}^2\right] - U(r) \tag{12.9.2}$$

The generalized coordinates are the distance r and angle θ. Since the angle θ does not appear in the Lagrangian, the associated momentum is conserved.

$$l = \frac{\partial L}{\partial \dot{\theta}} = \mu r^2 \dot{\theta} = \text{constant} \tag{12.9.3}$$

The quantity l is the magnitude of angular momentum vector **L**. Equation (12.9.3) implies that the angle changes monotonically because $\dot{\theta}$ never changes sign.

We observe that the kinetic energy has the form of equation (12.1.2).

$$T = \frac{1}{2}[\dot{r}, \dot{\theta}]\begin{bmatrix} \mu & 0 \\ 0 & \mu r^2 \end{bmatrix}\begin{bmatrix} \dot{r} \\ \dot{\theta} \end{bmatrix} = \frac{1}{2}\mu[\dot{r}, \dot{\theta}]\begin{bmatrix} \dot{r} \\ r^2\dot{\theta} \end{bmatrix} = \frac{1}{2}\mu\left[\dot{r}^2 + r^2\dot{\theta}^2\right]$$

The information in previous sections have demonstrated that these forms of kinetic and potential energy lead to conservation of energy. Hence $H = T + U = E$ and we write

$$H = E = \frac{1}{2}\mu\left[\dot{r}^2 + r^2\dot{\theta}^2\right] + U(r) = \text{constant} \tag{12.9.4}$$

Solving equation (12.9.3) for $\dot{\theta}$ and substituting into the Hamiltonian gives

$$H = E = \frac{1}{2}\mu\left[\dot{r}^2 + \frac{l^2}{\mu^2 r^2}\right] + U(r) = \text{constant} \tag{12.9.5}$$

Solving this equation for $\dot{r}$ produces

$$\frac{dr}{dt} = \dot{r} = \sqrt{\frac{2}{\mu}[E - U(r)] - \frac{l^2}{\mu^2 r^2}} \tag{12.9.6}$$

We now write equation (12.9.3) in differential form.

$$d\theta = \frac{l}{\mu r^2}dt, \quad dt = \frac{\mu r^2}{l}d\theta \tag{12.9.7}$$

12.9. CENTRAL FORCE PROBLEM

Integrating equation (12.9.6) produces a way to compute t. Using equation (12.9.7) in equation (12.9.6) and integrating gives θ.

$$t - t_0 = \text{sign}(\dot{r}) \int_{r(t_0)}^{r(t)} \frac{1}{\sqrt{\frac{2}{\mu}[E - U(r)] - \frac{l^2}{\mu^2 r^2}}} dr \qquad (12.9.8)$$

$$\theta(t) - \theta_0 = \text{sign}(\dot{r}) \int_{r(t_0)}^{r(t)} \frac{(l/r^2)}{\sqrt{2\mu[E - U(r)] - \frac{l^2}{r^2}}} dr \qquad (12.9.9)$$

Equation (12.9.8) gives r as an implicit function of t. Equation (12.9.9) gives θ as a function of r. These equations provide the complete solution to the problem.

The values of r which satisfy

$$U(r) + \frac{l^2}{2\mu r^2} = E$$

are called the turning points in the radial coordinate. $\dot{r} = 0$ at these locations and the radial velocity changes sign. On the other hand, the angular coordinate θ continually changes.

For a bounded motion between $r_{\min}$ and $r_{\max}$, we compute the change in angle $\Delta\theta$ for a complete radial cycle.

$$\Delta\theta = 2 \int_{r_{\min}}^{r_{\max}} \frac{(l/r^2)}{\sqrt{2\mu[E - U(r)] - \frac{l^2}{r^2}}} dr \qquad (12.9.10)$$

The paths are closed only if $\Delta\theta$ satisfies

$$\Delta\theta = \frac{2\pi m_{\text{rev}}}{n_{\text{per}}}, \qquad m_{\text{rev}}, n_{\text{per}} \in \mathbb{Z}$$

After n_{per} periods, the system completes m_{rev} revolutions and returns to the original position.

Consider a potential energy function of the form

$$U(r) = -\frac{k}{r} \tag{12.9.11}$$

where $k > 0$ is a constant. Gravity and electrostatic attraction of opposite charges have this form of potential. Equation (12.9.9) with this potential becomes

$$\theta(t) - \theta_0 = \text{sign}(\dot{r}) \int_{r(t_0)}^{r(t)} \frac{(l/r^2)}{\sqrt{2\mu \left[E + \frac{k}{r}\right] - \frac{l^2}{r^2}}} dr$$

If we choose $\theta_0 = 0$ for $t_0 = 0$ and $r(0) = r_{\min}$, then we obtain the solution

$$\cos\theta = \frac{\frac{l^2}{\mu k} \cdot \frac{1}{r} - 1}{\sqrt{1 + \frac{2El^2}{\mu k^2}}} \tag{12.9.12}$$

Arriving at this solution requires some clever techniques of integration and some details are given in [10]. Making the following substitutions simplifies the solution.

$$\alpha = \frac{l^2}{\mu k}, \qquad \varepsilon = \sqrt{1 + \frac{2El^2}{\mu k^2}}$$

Equation (12.9.12) can be written in a more convenient form.

$$\frac{\alpha}{r} = 1 + \varepsilon \cos\theta \tag{12.9.13}$$

This is the equation of a conic section with one focus at the origin, ε being the eccentricity, and 2α is the latus rectum.

Table 12.9.1 list all the possible trajectories. $V_{\min}$ is the minimum of the effective potential defined by

$$V(r) = -\frac{k}{r} + \frac{l^2}{2\mu r^2} \tag{12.9.14}$$

The trajectories describe the motion of the separation distance r in the center of mass coordinate system. They are not to be confused with actual orbits! When one of the masses is much greater than the other, then the heavier mass can be considered fixed relative to the other mass. The motion of the lighter mass around the heavy mass is described by the trajectories in table 12.9.1.

12.9. CENTRAL FORCE PROBLEM

Eccentricity	Energy Relation	Trajectory Shape
$\varepsilon > 1$	$E > 0$	Hyperbola
$\varepsilon = 1$	$E = 0$	Parabola
$0 < \varepsilon < 1$	$V_{\min} < E < 0$	Ellipse
$\varepsilon = 0$	$E = V_{\min}$	Circle
$\varepsilon < 0$	$E < V_{\min}$	N.A.

Table 12.9.1: Trajectories of motion

For elliptic orbits, the semi-major axis a and semi-minor axis b are given by

$$a = \frac{\alpha}{1 - \varepsilon^2} = \frac{k}{2|E|}$$

$$b = \frac{\alpha}{\sqrt{1 - \varepsilon^2}} = \frac{l}{\sqrt{2\mu |E|}}$$

These values are related to $r_{\min}$ and $r_{\max}$ using

$$r_{\min} = a(1 - \varepsilon) = \frac{\alpha}{1 + \varepsilon}$$

$$r_{\max} = a(1 + \varepsilon) = \frac{\alpha}{1 - \varepsilon}$$

These results can be used to derive Kepler's 3rd law.

$$\tau^2 = \frac{4\pi^2 \mu}{k} a^3 \qquad (12.9.15)$$

where τ is the orbital period. The derivation is found in [25].

12.10 Small Amplitude Oscillations

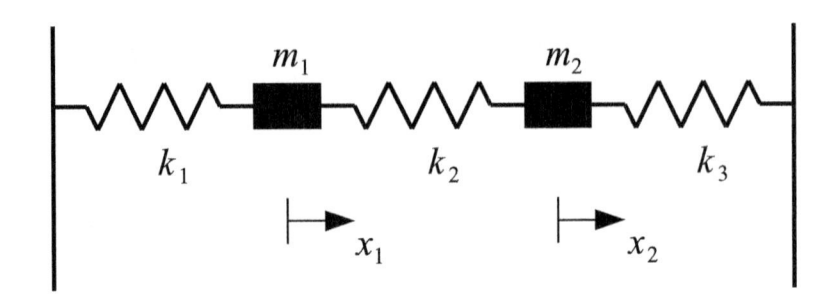

Figure 12.10.1: Example coupled mass-spring system

Figure 12.10.1 shows an example of a coupled mass-spring system. The two masses m_1 and m_2 are connected in the configuration by three springs with spring constants k_1, k_2, and k_3. This is a 2 DOF system with the generalized coordinates being the displacements x_1 and x_2. The displacements are measured relative to the equilibrium position of the center of mass of each object.

We are interested in generalizing the above system to n degrees of freedom with possibly non-cartesian coordinates $\mathbf{x} = [x_1, \ldots, x_n]^T$. Let $U(\mathbf{x})$ be the potential energy of the system. Let $\mathbf{u} = \mathbf{x} - \mathbf{x}_{\text{eq}}$ where $\mathbf{x}_{\text{eq}}$ is a stable equilibrium point. For small values of $\mathbf{u}$, we can use the 2nd order expansion to $U(\mathbf{x})$ at the equilibrium displacements $\mathbf{x}_{\text{eq}}$ and write

$$U(\mathbf{x}) = U(\mathbf{x}_{\text{eq}}) + \nabla U(\mathbf{x}_{\text{eq}}) \cdot \mathbf{u} + \frac{1}{2}\mathbf{u}^T \mathbf{H}(\mathbf{x}_{\text{eq}}) \mathbf{u} + r(\mathbf{u}) \qquad (12.10.1)$$

where $r(\mathbf{u})$ is $o(\|\mathbf{u}\|_2^2)$ as $\mathbf{u} \to \mathbf{0}$. The first term may be set to zero. This is due to the fact the only changes in potential energy affect the dynamics of the system and not the value itself. The gradient term is zero because $\nabla U(\mathbf{x}_{\text{eq}}) = \mathbf{0}$ is the definition of an equilibrium point. We ignore the error term because it is higher order. We assume that the Hessian term does not vanish and write the potential energy in terms of the displacements $\mathbf{u}$ from equilibrium. These will be the generalized coordinates in our formulation. The potential energy now becomes

$$U(\mathbf{u}) = \frac{1}{2}\mathbf{u}^T \mathbf{K} \mathbf{u} = \frac{1}{2}\sum_{i=1}^{n}\sum_{j=1}^{n} K_{ij} u_i u_j \qquad (12.10.2)$$

$$\mathbf{K} = \mathbf{H}(\mathbf{x}_{\text{eq}}), \qquad K_{ij} = \frac{\partial^2}{\partial x_i \partial x_j} U(\mathbf{x}_{\text{eq}}) \qquad (12.10.3)$$

12.10. SMALL AMPLITUDE OSCILLATIONS

The matrix $\mathbf{K}$ is symmetric positive definite because $\mathbf{x}_{eq}$ is a stable equilibrium point.

We assume that the kinetic energy has the form of equation (12.1.2).

$$T = \frac{1}{2}\dot{\mathbf{x}}^T \mathbf{A}(\mathbf{x})\dot{\mathbf{x}} = \frac{1}{2}\sum_{i=1}^{n}\sum_{j=1}^{n} A_{ij}(x_1,\ldots,x_n)\dot{x}_i \dot{x}_j \qquad (12.10.4)$$

where the matrix $\mathbf{A}$ is symmetric positive definite for all possible values of $\mathbf{x}$. We note that $\dot{\mathbf{u}} = \dot{\mathbf{x}}$ and write

$$T = \frac{1}{2}\dot{\mathbf{u}}^T \mathbf{A}(\mathbf{x})\dot{\mathbf{u}} = \frac{1}{2}\sum_{i=1}^{n}\sum_{j=1}^{n} A_{ij}(x_1,\ldots,x_n)\dot{u}_i \dot{u}_j$$

The elements of $\mathbf{A}$ are approximated by

$$A_{ij}(\mathbf{x}) = A_{ij}(\mathbf{x}_{eq}) + \nabla A_{ij}(\mathbf{x}_{eq}) \cdot \mathbf{u} + r(\mathbf{u}) \qquad (12.10.5)$$

$r(\mathbf{u})$ is $o(\|\mathbf{u}\|_2)$ as $\mathbf{u} \to \mathbf{0}$. The kinetic energy T is already a quadratic functional in $\dot{\mathbf{u}}$ and we approximate $A_{ij}(\mathbf{x})$ by $A_{ij}(\mathbf{x}_{eq})$.

$$T = \frac{1}{2}\dot{\mathbf{u}}^T \mathbf{M}\dot{\mathbf{u}} = \frac{1}{2}\sum_{i=1}^{n}\sum_{j=1}^{n} M_{ij}\dot{u}_i \dot{u}_j, \qquad M_{ij} = A_{ij}(\mathbf{x}_{eq}) \qquad (12.10.6)$$

The Lagrangian for the system in $\mathbf{u}$ coordinates is

$$L = \frac{1}{2}\dot{\mathbf{u}}^T \mathbf{M}\dot{\mathbf{u}} - \frac{1}{2}\mathbf{u}^T \mathbf{K}\mathbf{u} \qquad (12.10.7)$$

Lagrange's equations in vector form are

$$\frac{d}{dt}\frac{\partial L}{\partial \dot{\mathbf{u}}} - \frac{\partial L}{\partial \mathbf{u}} = \mathbf{0}$$

$$\frac{d}{dt}[\mathbf{M}\dot{\mathbf{u}}] + \mathbf{K}\mathbf{u} = \mathbf{0}$$

$$\mathbf{M}\ddot{\mathbf{u}} + \mathbf{K}\mathbf{u} = \mathbf{0}$$

The initial value problem is

$$\mathbf{M}\ddot{\mathbf{u}} + \mathbf{K}\mathbf{u} = \mathbf{0}, \qquad \mathbf{u}(t_0) = \mathbf{u}_0, \qquad \dot{\mathbf{u}}(t_0) = \dot{\mathbf{u}}_0 \qquad (12.10.8)$$

The initial value problem (12.10.8) may be solved by diagonalization. First, solve the general eigenvalue problem

$$\mathbf{K}\mathbf{v}_i = \omega_i^2 \mathbf{M}\mathbf{v}_i, \qquad i = 1, \ldots, n \qquad (12.10.9)$$

where ω_i are the general eigenvalues and $\mathbf{v}_i$ are general eigenvectors. The procedures to accomplish this task are given in [3, 4]. The eigenvectors are normalized through

$$\mathbf{v}_i := \frac{\mathbf{v}_i}{\sqrt{\mathbf{v}_i^T \mathbf{M} \mathbf{v}_i}}, \qquad i = 1, \ldots, n$$

The normalized eigenvectors satisfy the following properties:

$$\mathbf{v}_i^T \mathbf{M} \mathbf{v}_j = \delta_{ij}, \qquad \mathbf{v}_i^T \mathbf{K} \mathbf{v}_j = \omega_i^2 \delta_{ij}, \qquad i, j \in \{1, \ldots, n\}$$

Place the eigenvectors into the matrix

$$\mathbf{V} = [\mathbf{v}_1, \ldots, \mathbf{v}_n]$$

We have the following matrix diagonalizations:

$$\mathbf{V}^T \mathbf{M} \mathbf{V} = \mathbf{I}, \qquad \mathbf{V}^T \mathbf{K} \mathbf{V} = \mathbf{W} = \mathrm{diag}[\omega_1^2, \ldots, \omega_n^2]$$

where the second matrix is diagonal with ω_i^2 as its elements. The first equation implies that: $\mathbf{V}^{-1} = \mathbf{V}^T \mathbf{M}$.

Now let $\mathbf{u} = \mathbf{V}\mathbf{q}$ and hence $\mathbf{q} = \mathbf{V}^{-1}\mathbf{u} = \mathbf{V}^T \mathbf{M} \mathbf{u}$. Substituting into equation (12.10.8) gives

$$\mathbf{M}\ddot{\mathbf{u}} + \mathbf{K}\mathbf{u} = \mathbf{0}$$
$$\mathbf{M}\mathbf{V}\ddot{\mathbf{q}} + \mathbf{K}\mathbf{V}\mathbf{q} = \mathbf{0}$$

Now left multiply by $\mathbf{V}^T$ to get

$$\mathbf{V}^T \mathbf{M} \mathbf{V} \ddot{\mathbf{q}} + \mathbf{V}^T \mathbf{K} \mathbf{V} \mathbf{q} = \mathbf{0}$$
$$\ddot{\mathbf{q}} + \mathbf{W}\mathbf{q} = \mathbf{0}$$

We have the initial problem

$$\ddot{\mathbf{q}} + \mathbf{W}\mathbf{q} = \mathbf{0}, \qquad \mathbf{q}(t_0) = \mathbf{q}_0 = \mathbf{V}^T \mathbf{M} \mathbf{u}_0, \qquad \dot{\mathbf{q}}(t_0) = \dot{\mathbf{q}}_0 = \mathbf{V}^T \mathbf{M} \dot{\mathbf{u}}_0 \quad (12.10.10)$$

$\mathbf{q}$ is known as the *normal coordinates* for the system. Since $\mathbf{W}$ is a diagonal matrix, the equations decouple.

$$\ddot{q}_i + \omega_i^2 q_i = 0, \qquad q_i(t_0) = q_{i0}, \qquad \dot{q}_i(t_0) = \dot{q}_{i0}, \qquad i = 1, \ldots, n \qquad (12.10.11)$$

These equations are solved using the usual methods for 2nd order ordinary differential equations. We transform back to the original variables using: $\mathbf{u} = \mathbf{V}\mathbf{q}$.

Bibliography

[1] William E. Boyce and Richard C. DiPrima. *Elementary Differential Equations and Boundary Value Problems*. Wiley, New York, fourth edition, 1986.

[2] R. Creighton Buck. *Advanced Calculus*. International Series in Pure and Applied Mathematics. McGraw-Hill, New York, third edition, 1978.

[3] Anil K. Chopra. *Dynamics of Structures, Theory and Applications to Earthquake Engineering*. Pearson, Boston, fourth edition, 2012.

[4] Ray W. Clough and Joseph Penzien. *Dynamics of Structures*. McGraw-Hill, New York, second edition, 1993.

[5] Richard Courant. *Differential and Integral Calculus*, volume 1. Interscience, New York, second edition, 1936.

[6] Richard Courant. *Differential and Integral Calculus*, volume 2. Interscience, New York, second edition, 1936.

[7] Richard Courant and David Hilbert. *Methods of Mathematical Physics*, volume 1. Interscience, New York, 1937.

[8] Stephen H. Friedberg, Arnold J. Insel, and Lawrence E. Spence. *Linear Algebra*. Prentice Hall, Englewood Cliffs, New Jersey, second edition, 1989.

[9] I.M. Gelfand and S.V. Fomin. *Calculus of Variations*. Dover, Mineola, New York, 1963.

[10] Herbert Goldstein. *Classical Mechanics*. Addison-Wesley, Reading, Massachusetts, second edition, 1980.

[11] Aamer Haque. Green's functions for euler-bernoulli beams. 2015.

[12] Aamer Haque. Proofs of the energy theorems for beams and frames. 2017.

[13] Aamer Haque. Proofs of the energy theorems for elastic bars and trusses. 2017.

[14] R.C. Hibbeler. *Mechanics of Materials*. Prentice Hall, Upper Saddle River, New Jersey, seventh edition, 2008.

[15] Morris W. Hirsch and Stephen Smale. *Differential Equations, Dynamical Systems, and Linear Algebra*, volume 60 of *Pure and Applied Mathematics*. Academic Press, San Diego, 1974.

[16] Keith D. Hjelmstad. *Fundamentals of Structural Mechanics*. Springer, New York, second edition, 2005.

[17] A.N. Kolmogorov and S.V. Fomin. *Elements of the Theory of Functions and Functional Analysis*, volume 1. Graylock Press, Rochester, New York, 1957.

[18] A.N. Kolmogorov and S.V. Fomin. *Elements of the Theory of Functions and Functional Analysis*, volume 2. Graylock Press, Albany, New York, 1961.

[19] A.N. Kolmogorov and S.V. Fomin. *Introductory Real Analysis*. Dover, New York, 1970.

[20] Steen Krenk. *Mechanics and Analysis of Beams, Columns, and Cables*. Springer-Verlag, Berlin, second edition, 2001.

[21] Cornelius Lanczos. *The Variational Principles of Mechanics*. Dover, New York, fourth edition, 1970.

[22] L.D. Landau and E.M. Lifshitz. *Mechanics*, volume 1 of *Course of Theoretical Physics*. Pergamon Press, New York, third edition, 1976.

[23] Serge Lang. *Real and Functional Analysis*, volume 142 of *Graduate Texts in Mathematics*. Springer-Verlag, New York, third edition, 1993.

[24] Serge Lang. *Undergraduate Analysis*. Undergraduate Texts in Mathematics. Springer-Verlag, New York, second edition, 1997.

[25] Jerry B. Marion. *Classical Mechanics of Particles and Systems*. Harcourt Brace Jananovich, San Diego, second edition, 1970.

[26] Richard K. Miller and Anthony N. Michel. *Ordinary Differential Equations*. Academic Press, Orlando, Florida, 1982.

[27] Arch W. Naylor and George R. Sell. *Linear Operator Theory in Engineering and Science*, volume 40 of *Applied Mathematical Sciences*. Springer-Verlag, New York, 1982.

[28] M.H. Protter and C.B. Morrey. *A First Course in Real Analysis*. Undergraduate Texts in Mathematics. Springer-Verlag, New York, second edition, 1991.

BIBLIOGRAPHY

[29] Eduard Prugovečki. *Quantum Mechanics in Hilbert Space*. Pure and Applied Mathematics. Academic Press, New York, 1971.

[30] Robert Resnick and David Halliday. *Physics, Part I*. Wiley, New York, 1966.

[31] Robert D. Richtmyer. *Principles of Advanced Mathematics Physics*, volume 1 of *Texts and Monographs in Physics*. Springer-Verlag, Berlin, 1978.

[32] Walter Rudin. *Principles of Mathematical Analysis*. International Series in Pure and Applied Mathematics. McGraw-Hill, New York, third edition, 1976.